Behind the Screens

Winning the Ratings War at WAAY-TV

M.D. Smith, IV

Smith
BROADCASTING, INC.

Published by KDP Kindle Publishing Services

Typesetting and editing by CBA Publishing Services, LLC

ISBN: 978-1-7323849-3-4

Dedication

To the Smith Family before me, who pioneered the Radio & TV Broadcasting business.

To the Smith Family members, my sons and daughter, who worked in the TV business with me, and to ALL the great staff of people who came and went through the ranks and jobs at the TV station, and for them sharing their personal stories with me.

To all of them, and my wife Judy of 57 years, who put up with all this crazy, broadcasting stuff, this book is dedicated.

Table of Contents

Foreword

I met M.D. Smith, IV in 1973 when he hired me as WAAY's commercial producer. I'd not been particularly impressed by the facility during my one-day interview, but I did like M.D., as well as his father, M.D. Smith, III, and the job payed several thousand dollars more than I had been making at the CBS affiliate in Jackson, Mississippi, so I gladly took it.

A few weeks later, when my wife and I arrived in Huntsville, I turned on the TV in our motel to watch WAAY's news for the first time. It was a shock. I had expected the same slow moving, interview heavy, fare as Jackson. WAAY's news was anything but that. It was the best newscast I'd ever seen. Fast moving, full of video, even fun to watch, with great anchors. It was a revelation.

For the next five years I worked for and learned from M.D. Smith, IV. Three of the things I learned are with me to this day. They've influenced everything I've done professionally since then, so I'd like to share them here.

The only thing that matters in television is what's on the screen

Facilities are nice, the latest equipment is good to have, and big budgets are great, but none of that matters if the product on the screen fails. We had the tools we needed at WAAY, but rarely more, so we used everything to its fullest extent. We also had ingenuity that started at the top.

Winning is the most fun you will ever have in television

When I started at WAAY the market was dominated by arch rival WHNT. Their staff was arrogant and prideful, believing first place was their right. I watched the staff of WAAY change that. The pure fun of going from underdog to winner was a drug that still flows through my veins. It changed the direction of my career and has been a part of everything I've done since then.

Leadership is always the key to success

M.D. did not buy his way to victory. He earned it. Looking back, I realize our staff was united in one thing. We all wanted to beat WHNT and were determined to make that happen. That kind of unity is not accidental. I will go even further and say that WHNT lost their position because their leadership failed. Success or failure, every station is driven by leadership.

I've often thought it would be interesting to watch what career path M.D. Smith IV would have taken, had he had not grown up in a family business. I think he would have been in television, and like the rest of us, would have moved too many times to think about. I also think he would have ended up running a major station.

Hank Price
November 2018

Preface

Organized into five sections, this book begins with the Smith family's journey into broadcasting in 1928, the other businesses the family was involved in over the years and contains comments, illustrations and significant events that end with the writing and production of this book in 2018.

It includes many stories from the people who worked with the Smiths and WAAY-TV over their years of ownership. The second section of the book covers the beginning years of WAAY-31-TV in 1963-1965, which show the rudimentary equipment and methods required to get the news and programming *on the air* in those days. And, with a staff numbering less than twenty, everyone was multi-tasking before the term was ever invented.

The stories in section three are told by the people who were there, the "Alumni," all who had a story to tell; often amusing and surprising even including the author, M.D. Smith, himself. Many include illustrations and photos that will interest the reader and together work to show what went on *behind the TV screens* of the station that eventually won the ratings war and how it did so.

In the next portion of the book, the reader will share not only several collected memories of the author; you'll be a part of two real staff meetings exactly as they were delivered to the staff of 175 including research results and what they had to do to continue to win the "Ratings Wars" of the nineties.

Finally, the story of Channel 31 moves to 1995, a "peak" year with more projects, a live satellite truck & mobile studio, 24-hour news (by way of Live News Briefs *on the hour*-seven days a week) and concludes with a summary leading up to the sale of the station and reflections of the author who stays informed of the business from the General Managers who are still running the ship.

M.D. Smith, IV

Acknowledgements

All the past and present staff of WAAY-31-TV who participated in story contribution to this book.

To my author friends who helped with ideas and suggestions, Don Keith (N4KC ham call), Stephen Coleman, Mark and Laura Richerson and John E Carson. Also, Christine Brown and Anna Talyn Carson of CBA Publishing Services, LLC of Huntsville.

Tracy and Karin Slayton for title suggestions and the book cover art and layout.

And to the many others who helped proof and edit the manuscript.

I am most grateful.

Part 1

History of Smith Broadcasting

WBRC in Birmingham, WAAY, WNUE, WBYE, WJIG Radio Stations

WAAY-TV and HiWAAY Internet Services

Written by M.D. Smith, III, Kirby Jones Smith,

and M.D. Smith, IV

1925-1928

WBRC came to life on May 27, 1925 as a 50-watt AM radio station on 1210 KHz. The studio and transmitter, in the home of the owner, J.C. Bell and the call letters stood for Bell Radio Corporation WBRC, was the 2nd radio station on the air in Alabama. In the early days of radio, it was not uncommon for stations to be moved up and down the radio dial in an attempt to accommodate the increasing number of stations taking to the airwaves. In the next two years WBRC would increase its power to 250 watts (4/22/27), move to 1230 KHz (6/1/27), move to 1240 KHz (11/28/27) and move to 990 KHz on March 28, 1928.

M.D. Smith, Jr. bought WBRC, (Bell Radio Corporation = WBRC) in Birmingham, from J.C. Bell in 1928 for $2,000.00. The station operated four hours a day on (eventually on a frequency of 960 kc) with a power of 50 watts. The transmitter and studios were in the home of Bell, which was located in Fountain Heights.

(Mr. Bell's wife, Mary Emma Clayton Bell, did announcing (1925-1928) as was told in her death announcement of April 1, 2008. Yes, the date is correct, she was 105 years old, and was said to be one of the first "women disk jockeys." She lived in Vestavia.)

[Note, for a complete history of Channel 6, WBRC that includes history after the station was sold to Storer and is currently called "Fox6" in Birmingham.]

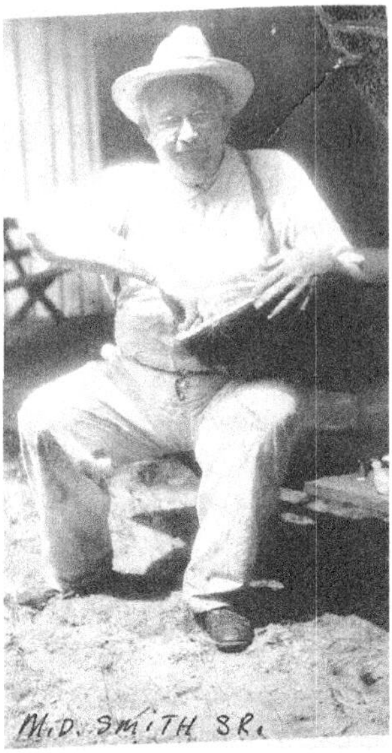

The father of M.D. Jr., M.D. Smith, Sr. was originally a minister and he was born in 1853. He had to make his own tent for a revival, made it better than any of that age, and eventually made some others and sold them. It was profitable, and he ended up going into that business full time as "Smith Mfg. Co." M.D. Sr. had two sons, M.D. Jr. (born 1890) and H.L. Smith came into the business Smith Manufacturing Company around 1908 in Dalton, GA. They made large and small tents. With his 2 sons M.D. Jr. and H.L., they started Awning and Tent companies in Birmingham, Atlanta and Chattanooga from their Dalton, GA home. Around 1910, M.D. Smith, Jr. came to Birmingham (probably with his father's guidance) and started Birmingham Awning and Tent Company, a branch of the Dalton Family business. M.D. Jr. married Eloise Hanye in February of 1912.

M.D. Jr. and H.L. had a disagreement and split the businesses in 1919, with M.D. Jr. taking Birmingham and Atlanta (later sold). There were plants then in Birmingham, Chattanooga and Dalton. MDS Jr. (born in 1890 and died in June 1937 from a nail wound in his arm from blood poisoning in days before Penicillin) ended up owning and running the Birmingham plant. H.L. took over Chattanooga and Dalton. By then, M.D. Sr. had died or retired. (Atlanta sold or closed down.)

Here is another photo of M.D. Smith, Jr. with his son, M.D. III (born in 1917) as a small boy.

1928-1929

The frequency was moved again on 11/11/28, this time to 930 KHz and the power was increased to 500 watts. In 1929 the transmitter was moved to a location behind the Birmingham Awning and Tent Works (the Smith's main business), at the corner of 12th Avenue and 27th Street North. Power stayed at 500 watts. Les Conners was hired as the Station's first salesman.

The studio was moved to the Old Athletic Club. Broadcast hours were extended to 12 hours a day when power was increased on April 30, 1929 to 1,000 watts daytime and 500 watts night.

1931

WBRC increased its power to 5000 watts and moved the studios to the Temple Theater on the mezzanine floor. The all-glass enclosure was known as The Crystal Studio. The transmitter was moved to a remote location in North Birmingham, known as Kilocycle, Alabama. Bill Young was hired as a second salesman and John Connerly was hired as chief announcer. (Don Connerly was an early part owner in beginning.)

The station was incorporated, and J.C. Bell bought 25 percent of the stock, Glenn Marshal bought 25 percent, and M.D, Smith, Jr and wife Eloise had 50 percent.

1932

Studio and offices were moved to the Bankhead Hotel. In December 1932, WBRC paid its first dividend of $5,000 to its stockholders. Radio had become a profitable business. Mr. Smith, Jr. continued his Birmingham Awning and Tent business and MDS, III was 14 and working part time in the Awning and Tent Company. He went into the business full time a few years later.

1935

Due to the rapid expansion of the radio business, WBRC moved to their present location on the corner of 19th Street and 2nd Avenue North. The staff had increased to 20 and WBRC became affiliated with the NBC network.

1937

M.D. Smith, Jr. died in 1937 at the age of 46 from blood poisoning he got from a cut. (Born May 24, 1890 – Died June 9, 1937)

His wife, Eloise Hanye Smith (later Eloise Smith Hanna) took control of Bell Radio Corporation (WBRC). (Born Apr. 7, 1893 – Died Dec. 11, 1980)

M.D. Smith, III and Elizabeth Kirby Jones married August 4, 1937. Because of the death of his father, the wedding was smaller than planned and held at home. MDS, III continued to run Birmingham Awning & Tent Company while his mother, Eloise, took over the radio business.

1938

The present transmitter location, at Finley Blvd and Highway 78 was put in operation on February 3, 1938 when the station's power was 5,000 watts daytime (the current power level) and night time raised to 1,000 watts. On October 1, 1940 a second tower was added to make the station directional to the south at night. On March 29, 1941 WBRC was ordered by the FCC to move their frequency to 960 KHz where it remains today.

(Of interest, the layout of this building with offices to the left, entrance hall & live studio in center, and on the right the announcer/control room and rear of that was the transmitter room. MDS, III built that exact same layout in 1955 when he built WBYE-AM in Calera, Alabama as the first radio station he would build and own himself after Eloise sold WBRC-AM-FM & TV in 1953. The WBYE building was still the original in 2017 when I visited.)

1939

Eloise Smith bought out Glenn Marshal's stock for $25,000 and took over the reins of the Corporation as President and General Manager. In 1940, J.C. Bell died and Eloise Smith purchased Mrs. Bell's stock for $35,000.

In 1940, Eloise married Dr. Hanna and changed her legal name to Eloise H. Hanna.

1940

December 6, 1940 *M.D. Smith, IV* was born. (Pearl Harbor would be bombed a year and a day later.)

1941-1945

During the war years, Eloise H. Hanna joined the Motor Corps and rendered valuable service transporting our troops to and from arrival and departure points. She was also a member of the American Red Cross.

During the war years, there was much talk about FM radio. It was believed that after the war, all AM stations would convert to FM.

WAR YEARS

added by *Kirby Jones Smith* (Mrs. M.D. Smith, III)

M.D. Smith, III was in business as Birmingham Awning & Tent Co. After Pearl Harbor (December 7, 1941), the Quartermaster came to see us about war work, since we made tents, awnings, and trampolines at Birmingham Awning and Tent, and practically said we had to go in war work. We made everything from parachutes to underwear, as well as tents, during the war.

We then formed Marson and Company—a 50-50 partnership, M.D. Smith III and Kirby J. Smith—we were known as in the needle trade. They loaned us $50,000 and furnished all material. We paid back the $50,000 after the first item. We started making bandoleers (shell clips) at our location at 213 South 21st Street, Birmingham, Alabama. We bought 40 sewing machines, a cutting table, and turned out 5 million bandoleers.

During that time, the Quartermaster seemed pleased and came to us and asked us to make parachutes. They were all white and all shroud lines too. We had a time as they were all cut on bias to center top ring. There was always a government inspection in the three plants, to even count the stitches as per specs. After the first 425 came off and were inspected, we finally had one accepted. I'll never forget MD III held it up and said we had conquered the order to parachutes. We went on and made and shipped a total of 75,000 parachutes. At a price we received $50.00 a parachute, we made a ton from them (MD IV was 1 year and 3 months old in March of 1942. I'll never forget MD III said to MD IV, we are on our way!)

We were also awarded contracts for 100,000 barrack bags, men's shirts, general issue, and jungle hammocks, as the war was invading through the jungle, through the Philippines. They were very complicated, made to hang between two trees, and had hammock canvas bed with sides of netting, a zipper on all sides, and a waterproof top. We made 25,000 jungle hammocks and were paid late, as not many plants in the United States would undertake them. We had then bought two other buildings, each two floors, to make these items. We worked three shifts at all plants with 800 people (seamstresses) and long cutting tables in each plant. They gave us the patterns of heavy-like Formica. We laid 50 layers for each cut. The minimum wage was 40 cents an hour.

Everything was rationed and every month, every adult received a book (similar to food stamps) for gasoline, meat, sugar, alcohol (one pint per person a month).

The day the Armistice was signed, every war contractor received a wire to cease. They sent 18-wheel trucks to collect all materials. No matter what stage the item was in, they picked it up.

1945

We sold most of the commercial sewing machines but kept some and when white material was starting to be made again, we bought white broadcloth and made 1,000 white shirts for Arrow. We paid for cloth, labor, thread, buttons, and everything. We did make a little but decided MD III would go back to WBRC Radio. During the war days, I made $500 a month and MD III made $1000 a month. We had a big savings account. When he went back to WBRC, I stayed and refurbished Birmingham Awning (Now MARSON) and ran it. The "MAR" was from the Marcellus and the "SON" was from Davidson (what M.D. is the abbreviation for). The logo had an orange sun over the name.

Here is a photo of the Marson building on 213, South 21st Street around 1945. Notice the nice cars parked in the alley beside the building.

1946

Anita was born the first year back in the normal business, so we sold Birmingham Awning and Tent and Venetian Blind to Guarantee Awning in Birmingham for $7500. I had a nurse at home and I had a couple (cook and chauffeur) on place all time. You see why MD III gave me a 1945 yellow Cadillac when Anita was born.

(End of Mrs. Smith III's comments)

1946

Mrs. Hanna applied for, and received, a construction permit to put WBRC-FM on the air. It was the **world's most powerful FM station**, operating with a power of **500,000 watts**. During 1947, her son, M.D. Smith, III joined WBRC-AM and FM as salesman. Smith was later promoted to Program Director and Vice President of WBRC.

1948

FM radio did not sweep the country as expected, and due to large losses, the FM station was taken off the air in June of 1948. Mrs. Hanna then turned her attention to the emerging medium of television.

Many felt that TV, like FM, was just a flash in the pan and would never be anything more than just a novelty. Mrs. Eloise Hanna had never lost her pioneering spirit though, so she borrowed $150,000 and building began. The original FM building with the massive self-supporting tower, was turned into the TV transmitter and very small studio was made on the basement floor of the existing building. So, other than the TV equipment, everything else was in place on Red Mountain. WBRC-TV went on the air on July 4, 1949. It was put IN the FM building. The photo of Eloise Hanna at her desk shows the nameplate mis-spelled "Elose", but it was an expensive gift and she chose to leave it that way.

1949

July 4, 1949 a yearbook shows the details of Channel 4 with the Dumont Television Network. Channel 4 had and original power of 15KW Visual and MDS III as the only contact as PD. In early 1953, the FCC made them change to Channel 6 to avoid interference with Nashville. It cost a lot to modify the transmitter to change channels, but it had to be done.

BIRMINGHAM

WAPI 1922	1070 5,000 DA-N	Voice of Alabama Inc. 701 Protective Life Bldg. 3-8116	CBS Radio Sales	Ed Norton, chmn. of bd. Thad Holt, p & gm C. P. Persons Jr., cm Lionel Baxter, m of opers., pmm & nd	James L. Evans, ce E. H. Mitchell Jr., t John McCune, fd
WAFM(FM) 1947	*Ch. 258 (B)* *99.5mc* *515kw*	*(WAPI, AM Affiliate)*		*Norman S. Hurley, ce*	
WAFM-TV 1949	Ch. 13 210-216mc 13.6kw-aur.; 26kw-vis.	(WAPI, AM Affiliate)	ABC CBS		
WBRC 1925	960 5,000 DA-N	Birmingham Bcstg. Co. 1727½ 2nd Ave. N. 4-7741	NBC Raymer	Eloise S. Hanna, o & p G. P. Hamann, gm & tech. d D. D. Campbell, cm W. H. McCain, pd & nd	J. T. Orr, pmm Don Norman, fd Richard Hawley, sd Mary Badham, wd
WBRC-TV 1949	Ch. 4 66-72mc 7.5kw-aur.; 15kw-vis.		DuMont	M. D. Smith III, pd	

When WBRC TV went on the air, there were only 12 TV sets in Birmingham, and they were in dealers' show windows. M.D. Smith, III was appointed as Operations Manager of TV. Oliver Naylor was salesman; Evelyn Allen, traffic; Bob Farris, film editor; Felix Lewis, cameraman; and Hardy Carl, chief engineer. Nora Harrimontrec was artist. So, with a staff of seven, they were off and running.

They were affiliated with the DuMont & NBC Television Networks. However, there was no cable connecting the TV station with New York, so all network programs came to us on kinescope or film made from a TV screen. The quality was quite poor.

M.D. Smith, III and Mrs. Hanna continued to add equipment and personnel to the TV station. Live studio cameras were added in 1950, and by 1953 the staff had grown to 15. Today WBRC TV, Channel 6, employs more than 150 people.

1951

Fall, 1951 [*Note from M.D. Smith, IV - My father told me about how creative they had to be, even decorating the small studio with toilet tissue for background effects, since there was virtually no budget for props.*]

So here is that creativity in action with the costumes of the staff of WBRC-TV (and probably radio also) at a station party around Halloween of 1951. Of this group of five, M.D. Smith, III is on the left in the dress with the bandage on his leg from a serious auto accident in early 40s and a scar that was on bone. And finally, the whole staff and all their costumes. TV folks have always known how to have fun.

1952

1952

INGHAM NEWS

Mrs. Eloise Hanna will tour Europe with press group

Mrs. Eloise Smith Hanna, president of Radio Stations WBRC and WBRC-TV, will tour Europe and the Middle East next month.

She is among a group of press and radio executives who will make the tour sponsored by the U. S. State Department.

Mrs. Hanna will leave New York Saturday.

Approximately 50 editors and executives are included in the group.

The tour will include visits to Berlin, Frankfurt, Bonn, Vienna, Belgrade, Istanbul, Ankara, Teheran, Tel Aviv, Jerusalem, Cairo, Athens, Rome, Paris and London. The group is scheduled to return to the states on April 1.

* * *

MRS. HANNA WILL participate in conferences with England's Lord Beaverbrook and Prime Minister Winston Churchill, Gen. Dwight D. Eisenhower, Marshal Tito, Pope Pius, President Auriol of France and other officials of European and Middle Eastern states.

They are scheduled to sit in on a presidential press conference before leaving and hear a panel discussion in which participants will be Sen. Brewster (R., Me.), Sen. Morse (R., Ore.), Sen. Moody (D., Mich.), and Sen. Smith (D., N. C.)

Also scheduled for the group is another discussion period with Dr. George Denny Jr. as moderator. Speakers will be Commentators Dorothy Thompson, Henry Hazlitt, Leo Cherne and Lester Markel.

Plans tour—Mrs. Eloise Smith Hanna.

A newspaper article & photo appeared on Mrs. Hanna and a visit to Europe and the Middle East she made with a Press Group sponsored by the U.S. State Department.

Election Returns of 1952 when Eisenhower was elected. Note the NBC, Channel 4 logo in the top of the photo. The station changed to Channel 6 shortly after this night. (Early 1953) The FCC said there was too much overlap with Ch. 4 in Nashville, (making viewing in Huntsville very difficult) so the FCC ordered WBRC to switch over to Channel 6. There was a fair amount of expense doing that, but it had to be done.

Channel 4 Election Returns just after getting live cameras at the station, November of 1952. Kirby Smith (Mrs. M.D.) can be seen on left. She is woman in dark hair just under man holding sign.

1953

In this year, Mrs. Hanna decided to retire and sold WBRC TV/AM-FM to Storer Broadcasting for $2.3 million.

(Comment by *Kirby Jones Smith*) After a year that sale, we started our own broadcast station and formed a corporation.

Written by M.D. Smith, IV from this point on, March 1992

1955

M.D. Smith, III and wife Kirby Jones Smith, put radio station WBYE AM on the air in Calera, Alabama. Built from scratch, on 1370 kc with a power of 500 watts, daytime-only license. One of the sales mottos was, "Your Best Buy By Far Is Bye." It was the first top 40 radio station in the area. M. D. Smith, IV worked one summer in 1956 at the station as music librarian, assistant book keeper and remote technician (set up remote mics for Sunday Church remote broadcasts).

The station is still on the air and this photo of was taken in July of 2002. The building is exactly the same size and layout as when it was first built, on the road out of Calera on the way to Montevallo, Alabama. It was built on the same floor plan as WBRC's original transmitter/studio building on Finley Blvd.

Here is M.D. Smith, IV standing under sign on 7-27-02 and another of the road sign on the highway.

1956

M.D. Smith III bought WJIG AM in Tullahoma, TN, already on the air on 740 kc with a power of 1000 watts daytime only. Henry Beam, who worked as a salesman for a beer company, became General Manager/Sales Manager. M.D. Smith, III resigned from Storer Broadcasting to devote full time to the radio business and the two stations.

1958

M.D. Smith III buys WHBS AM, 1550 AM, (5,000 watts day and 500 watts, directional antenna at night) from the Huntsville Times, changes call to WAAY, and changes format from Adult Listening music to Rock and Roll Top 40. A petition is circulated to boycott the radio station and run new sound out of town. Several employees quit that could not tolerate Rock and Roll music. Dick Essner becomes Chief Engineer as well as Disc Jockey. M.D. Smith IV, who has been working as Record Librarian, also began some Disc Jockey duties in the summer of 1958, filling vacation shifts.

WAAY 1550-HUNTSVILLE *Radio*

WAYS OFFICIAL
FAMILY SURVEY NUMBER 31
WEEK OF AUGUST 3, 1959

1. A BIG HUNK OF LOVE/MY WISH COME TRUE — ELVIS PRESLEY — RCA
2. TIGER — FABIAN — CHANCELLOR
3. CRACKIN' UP — BO DIDLEY — CHECKER
4. JUST A LITTLE TOO MUCH/SWEETER THAN YOU — RICKEY NELSON — IMPERIAL
5. WHAT IS LOVE — THE PLAYMATES — ROULETTE
6. SINCE YOU'VE BEEN GONE — CLYDE McPHATTER — ATLANTIC
7. LONELY BOY — PAUL ANKA — ABC PARAMOUNT
8. SEA OF LOVE — PHIL PHILLIPS — MERCURY
9. RAGTIME COWBOY JOE — DAVID SEVILLE & CHIPMONKS — LIBERTY
10. I'LL BE SATISFIED — JACKIE WILSON — BRUNSWICK
11. LAVENDER BLUE — SAMMY TURNER — BIG TOP
12. WHAT A DIFFERENCE A DAY MAKES — DINAH WASHINGTON — MERCURY
13. DANNY BOY — SIL AUSTIN — MERCURY
14. HIGH HOPES/ALL MY TOMORROWS — FRANK SINATRA — CAPITOL
15. PERSONALITY — LLOYD PRICE — ABC PARAMOUNT
16. THERE GOES MY BABY — THE DRIFTERS — ATLANTIC
17. SWEET SUGAR LIPS — THE KALIN TWINS — DECCA
18. I'M COMING HOME — MARV JOHNSON — UNITED ARTIST
19. SINNER MAN — TOMMY SANDS — CAPITOL
20. THE WONDER OF YOU — RAY PETERSON — RCA
21. PURE LOVE — SONNY JAMES — CAPITAL
22. SMALL WORLD — JOHNNY MATHIS — COLUMBIA
23. CROSSFIRE — JOHNNY & THE HURRICANES — WARWICK
24. A LONG WAY FROM ST. LOUIS — PEGGY LEE — CAPITOL
25. PARTNERS — JIM REEVES — RCA
26. STAY — DEE ARLEN — COLUMBIA
27. YOU ARE IN LOVE — PERRY COMO — RCA
28. MY HEART IS AN OPEN BOOK — CARL DOBKINS JR. — DECCA
29. THIS I SWEAR — THE SKYLINERS — CALICO
30. SEE YOU IN SEPTEMBER — THE TEMPOS
31. SOFT SUMMER BREEZE — EDDY HEYWOOD — MERCURY
32. THE WAY I WALK — JACK SCOTT — CARLTON
33. I'M GONNA BE A WHEEL SOMEDAY/I WANT TO WALK YOU HOME — FATS DOMINO — IMPERIAL
34. HUSH-A-BYE — THE MYSTICS — LAURIE
35. CHOSEN FEW — FOUR LADS — COLUMBIA
36. THANK YOU PRETTY BABY/WITH ALL MY HEART — BROOK BENTON — MERCURY
37. FIVE FEET HIGH AND RISIN'/I GOT STRIPES — JOHNNY CASH — ATLANTIC
38. IF YOU LOVE ME/SO HIGH SO LOW — LAVERN BAKER — ATLANTIC
39. I'VE BEEN THERE — TOMMY EDWARDS — MGM
40. CRY — KNIGHTSBRIDGE STRINGS — TOP RANK
41. BLUES STAY AWAY FROM ME — BILLY VAUGHN — DOT
42. HEY LITTLE GIRL/A GIRL LIKE YOU — GARY STITES — CARLTON
43. ONLY WISHFUL THINKING — PETER KIM — CLICK
44. TIME MARCHES ON — ROY HAMILTON — EPIC
45. BIG SURPRISE — FOUR PREPS — CAPITOL
46. SOLDERS JOY — JERRY REED — NRC
47. TWICE AS NICE — CLYDE McPHATTER — MGM
48. SLOW MOTION — WADE FLEMMONS — VEE JAY
49. LOTS OF LUCK — THE HILLTOPPERS — DOT
50. THIS LITTLE LOVE OF MINE — ED TOWNSEND — CAPITOL

51. WHAT'D I SAY (Part 1&2) — RAY CHARLES — ATLANTIC
52. LINDA LOU/RED SAILS IN THE SUNSET — RAY SHARPE — JAMIE
53. JOHNNY ROB — JOHNNY HORTON — COLUMBIA
54. THE MUMMY — BOB McFADDEN AND DOR
55. CHECK-MATED AND BINGOED — BILLY AND DON HEART — ROULETTE
56. THE THREE BELLS — THE BROWNS — RCA
57. I'VE COME OF AGE — BILLY STORM — COLUMBIA
58. WITH MY EYES WIDE OPEN — PATTI PAGE — MERCURY
59. WE BELONG (TOGETHER) — JIMMY FORD — STYLO
60. KISSIN TIME — BOBBY RYDELL — MGM

"WAY" STARLIGHTERS
SPOTLIGHT SONG HITS

DAN AKENS — THE SHAPE I'M IN/YAYA — JOHNNY RESTIVO — RCA

DICK ESSNER — WALKING ON AIR — JIMMY BOWEN — ROULETTE

JERRY LEE — I'M GONNA GET MARRIED — LLOYD PRICE — ABC PARAMOUNT

DEE SCOTT — TALL COOL ONE — THE WAILERS — GOLDEN CREST

1550 YOUR RADIO "WAY" IN HUNTSVILLE

DAN AKENS
THE SHAPE I'M IN/YAYA
JOHNNY RESTIVO — RCA

DICK ESSNER
WALKING ON AIR
JIMMY BOWEN — ROULETTE

JERRY LEE
I'M GONNA GET MARRIED
LLOYD PRICE — ABC PARAMOUNT

DEE SCOTT
TALL COOL ONE
THE WAILERS — GOLDEN CREST

Here's one of the weekly "Family Survey Top 60" record lists of August 1959. A bit closer shot shows the four DJ's pick hits of the week. The DJ's are Dan Akens, Dick Essner, Jerry Lee (aka Jerome Tisdale) and Dee Scott (aka M.D. Smith, IV). This was list #31 for the 31st week of the year. This week, Dee Scott's Pick Hit, Dee Scott's Pick Hit of the week was "Tall Cool One" by The Wailers on Golden Crest records.

DEE SCOTT
TALL COOL ONE
THE WAILERS — GOLDEN CREST

At WAAY Radio, commercials were played on a pair of Magnecord PT-6 reel to reel, recorders. Each commercial, jingle and promo was on a 3" reel of tape and running multiple spots required great dexterity to rewind and rethread one recorder while the other was playing. The rewind motors would break a finger if it got in the spokes of the tape hub during rewind. You had spare heavy-duty leader tape, a tape splicer and splicing tape to continue to replace the leader that would whip off sometimes during rewind or the fast threading you had to do. Spotmaster cartridge tapes that came in later years were fantastic.

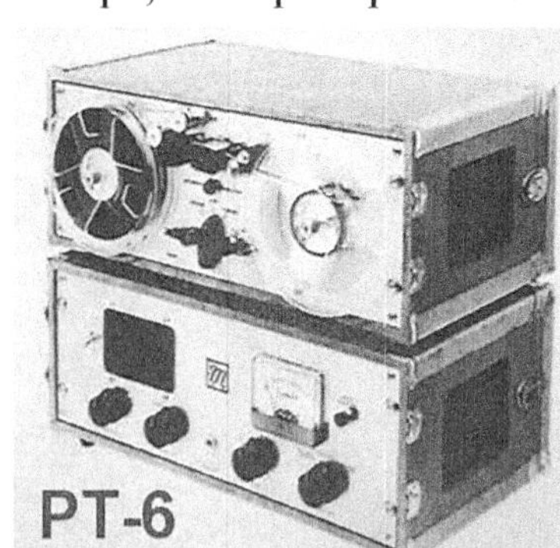

1959-1960

M.D. Smith, III, at the strong urging of his wife, Kirby Jones Smith, buys WFBS (Fort's Best Station) in Ft. Walton Beach, FL from French Brown. The station had gone on the air in Ft. Walton during the summer of 1956 on 950 kcs at 500 watts, daytime only. FCC approval of the sale came in May of 1960 and MDS III changed call letters to WNUE, the "Nue" station in town with the Top 40 format that's always "Nue". Here's a photo of the first coffee cup ever made. It says, "For Better Business Tomorrow, WNUE Today." Station was on 950 Kcs daytime. By changing frequency to 1400, station was able to go night-time and broadcast 24 hours a day. Power was 1,000 watts day and 250 watts night. Kirby Smith also is involved with the bookkeeping, accounting, record list, music, contests and promotions. This is the original building before the major front addition, block lattice wall and call sign in giant letters on the wall.

1963

M.D. Smith, IV graduates from college in June. In college he worked for WNPT and WJRD as a first-phone engineer. He also worked for the student AM station WABP as the Sales Manager. This is his first Business Card.

Begins to work for WAAY-Radio as Merchandising Director, engineer and disc jockey. We are planning to put a TV station on the air on Channel 25.

Charles Grisham plans to put one on the air on Channel 19. The lone station, WAFG-TV, Channel 31, which has been struggling, decides to get out of the business faced with the impending competition (Sign on Car for Ch. 25) and in the Summer of 1963 offers to sell it to Smith Broadcasting for $500,000.00 That is about what was planned to be spent on Channel 25, so it is decided to take their offer. Jim Beasley from Sweet Sue in Athens is the majority stockholder who confirmed the sale. The minor stockholders in Huntsville did not want to sell but had no choice. The deal was concluded in July and we apply to the FCC for an ownership change. Meanwhile, Channel 19 begins construction on property next to channel 31. Channel 25 is later awarded to the Alabama Educational Television Authority.

Robert (Cactus) Gay, age 23, is one of the engineers employed by WAFG-TV when it is bought by Smith and stays with the company until the present time. The station was broadcasting on a 125-foot tower with 240,000 watts.

Call letters are changed to WAAY TV.

First mascot was a beagle hound.

November - The FCC finally approves the sale and the Smiths take over Channel 31 with a staff of 17 people. M.D. Smith, IV is Operations Manager. Maury Farrell is Sales Manager and MDS, III is a remote General Manager. Channel 31 WAAY-TV is affiliated with the ABC television network. Channel 19 (WHNT) also goes on the air in November with a 250-foot tower and 600,000 watts and the CBS TV network. Channel 31 had ONE Video Tape Machine. It was an Ampex, VR-1000. It used 2" tape. It was monochrome (or course). It took up a space of 4 feet by 4 feet by 6 feet and that did NOT include the TWO 6-foot-high equipment racks of electronics (all tubes) to run the machine. It had massive power supplies and tubes had to be changed every day to keep it running. Only one spot on a break could be on Video Tape since there was no way to splice or edit tapes together. The other commercials had to be Slides and Audio Tape. A picture of Master Control with projectors and slide chain in the foreground and the operator (Lamar Gilbert) at the "switcher" in the background.

When we bought the station, they were reading all commercials LIVE from the announce booth. The first thing we did was put in Audio Cart Machines with trip cues for the slides to eliminate the announcer and the switcher didn't have to read along with the copy and trip the slides at the right time. There was a Live Romper Room show every weekday morning. It was the major production of the day. We used two,

mismatched black & white studio cameras. One was a General Electric Image Orthicon camera with a 3-lens turret and the other was a General Electric Videcon camera with a zoom lens. We liked the zoom lens a lot and eventually replaced the turret with another very expensive "Zoomar" zoom lens (as seen in photo above) we bought used at a good price. In the photo, M.D. Smith, III is running Camera 3.

The slides for commercials were shot on black & white "reversal" film. When it was processed it was a slide ready to show, rather than a negative to make a print from. That station had a very small, old B/W film processor that processed 16mm B/W film for the nightly newscast at 6 and 10. It ran slowly at 4 feet per minute and it took 100 feet of film for less than 3 minutes of film, much of which would be edited out and thrown away. However, we tried to only shoot what we wanted to use to save on film costs. If a photographer could get 2 or 3 stories on one 100' roll of film, we thought he was great. Our 6 and 10 P.M. News Casts were called "The WAAY of the World." The station also owned one OLD Auricon sound 16mm camera. It was an optical sound track and the electronics (tubes) used numerous batteries for the filament voltage, plate voltage and motor power.

Late November - Just two weeks later, President Kennedy is shot and ALL TV stations cease broadcasting commercials for 5 days. This is tough for a struggling new TV station.

WAAY Radio, meanwhile, is making a lot of money as the dominant radio station in Huntsville and the revenue is used to pay the losses at channel 31 during these first few struggling years.

1964

January–WAAY-TV broadcasts the first of the Annual C-P Telethons from our studios. Running the camera is M.D. Smith, III and in front of it is Maury Farrell.

"Cactus" Gay takes a turn on camera, as does almost everyone in the station during this event.

1965

March - M.D. Smith, IV AND M.D. Smith, III both get pilot's licenses. The first aircraft owned by WAAY Radio and TV was a Cessna 172 Skyhawk. Here is a photo of the tail of that aircraft with the "News Hound" logo that has been in use at the radio stations for 9 years. Aerial film of tornado damage in parts of Western Alabama are shot later that month.

Here is an aerial photo of WAAY-Radio made from the airplane at about the same time. You see the three-tower directional array maintained by Chief Engineer Robert A. "Cactus" Gay. The center tower is self-supporting and was used originally when the station was daytime only. The other two are guyed and were put up to form the directional array to allow night time power and 24-hour operation.

August - Anita Smith and Wayne Johnson are married. Wayne goes to work at WAAY Radio in June of 1966 after getting a degree in Marketing at U of A.

1967

Channel 31 builds a new 400-foot tower (the old was 125 feet tall) and increases power from 240,000 watts to 1,240,000 watts in order to serve Florence. With this tower and power improvement we are able to

convince NBC we would be a far better affiliate to have than WMSL, channel 23 located in Decatur. The Network agrees. In January 1968, Channel 31 switches to NBC to improve ratings. Maury Farrell and M.D. Smith, IV have picture taken in front of a Giant Floral Peacock in the studio.

M.D. Smith, III heard a song in another town, and found out that the music was available for purchase and a special pressing of a 45 RPM record. The company would either write your song lyrics or you could do them yourself. Mr. Smith decided to have the record made for Huntsville and Ft. Walton Beach (Home of WNUE Radio). He wrote the lyrics for both songs and the TV station pilot (M.D. Smith, IV) flew the company airplane for the aerial shots of the two cities. A month later, "Huntsville, My Home Town" was born. So was the same song with different words, both written by M. D. Smith, III for WNUE in Ft. Walton was made.

August - After both Smith's get Multi-Engine ratings, the station converts to its first of four Piper Aztec Twin Engine (250 horsepower each engine) airplanes. This is the Piper with the tail painted with call signs and the special assigned number with the FAA of N31TV. It is 60 miles an hour faster than the Cessna 172.

1968

February - Channel 31 Signs an agreement with Jackson Way Baptist Church to begin televising the Sunday worship service with the first LIVE microwave remote in the valley. Maury Farrell and M.D. Smith, IV look on as Rev. Jim Dorriety signs the agreement. The broadcasts begin in June of 1968.

April - Channel 31 converts to Color Video Tape and Color studio cameras. Having bought a used color film-chain/island from WTOP in Washington, DC in 1965 and the first station to broadcasting anything local in color. The two new 2", quad rotating head tape recorders (RCA TR-70 and an RCA TR-22 (used but "high-banded" for color)) were a major improvement. Before we only had one black & white Ampex VR-1000, all tube video tape recorder. It was a monster and good to retire it. It sat in storage for a year with no one wanting it and was finally sold for a few hundred dollars as parts. The big RCA TK-42 Color Studio cameras were huge. They each had one 3" Image Orthicon

tube for the picture and three 1" vidicon tubes for the 3 color guns. The cameras cost $50,000 each. A complete set of replacement tubes was $12,000 per camera. As the first station to broadcast live newscasts in color, we were very proud. Channel 19 had a ratings edge over us, even in B/W, but they also bought new color cameras about 8 months later.

1968

August - After 6 years working for WNUE (1 yr.) and WAAY (5 yrs.) radio, Dan Whitsett, takes Maury Farrell's place as General Sales Manager of Ch. 31. At that time the TV station was gross billing about $300,000 a year with about 25 employees.

1968

ABC signs up with WMSL, Channel 23 who get a frequency (Channel 48) moved to Huntsville and move from channel 23 in Decatur to channel 48 in Huntsville. They build a 900-foot tower but are on a mountain that is 500 feet lower than Monte Sano, so their antenna is not that high above average terrain. Maury Farrell is their new general manager working for Frank Wisnett of Decatur. Their new call letters are WYUR ("Your" TV station). They are bought years later by American Family Life Insurance company and change calls again to WAFF (American Family's Finest).

1969

The late 60's and early 70's PAMS jingles packages arrived. These were the latest and most expensive of the jingle packages, and both M.D. III and M.D. IV flew to Dallas to sit in a booth during the recording sessions and change the lyrics as they were being made if something didn't sound just right. The music was pre-recorded, and the singers just sang along with the music track, so re-takes were easy.

"Fifteen-Fifty . . . W - A - A - YYYYYYYY!"

1970

Channel 31 decided to convert from B/W film for the TV news to Kodak's new color news film. We install a high-speed color processor and tanks of color chemicals. The old processor operated at 4 feet of film per minute. We felt the new 20 foot per minute processor was a speed demon. It certainly used up a lot of color film at $10.00 per 100-foot roll (about 2 1/2 minutes of film). We processed and used about 10 rolls per day for the newscasts. This was a major expense to our station (about $5,000 a month).

1971

January - For a year, M.D. Smith, IV was the 6 & 10 p.m. weatherman. The maps in those days were painted on canvas with a Plexiglas cover. The daily weather was drawn with "paint pens" of various ink colors. It had to be removed with lacquer thinner. The temperature boxes and the forecast was "menu board" with plastic stick-on letters. The nice part was that you could draw right on the map to explain what was happening.

March - WAAY Radio and WAAY-TV do our first major joint promotion. The Bridal Fair held at the Sheraton Motel on University Drive. It was a big success and made money for both stations, but it was a real problem deciding which salesman at radio or TV got how much commission from selling it. This and other logistical problems prevented many other joint promotions of this nature in the future.

1972

March - A newly designed News Set appeared. The station's first true Meteorologist, Walton Jones joins the team of Adrian Gibson, Sam Depino and Ed Sisson. The weather maps are the same material but get reversed.

1973

March - The anchor team grows. Dan Jamison, Sports Director is added to the team (back left), along with Walton Jones (back right) and (L to R) Ed Sisson, Adrian Gibson and Sam Depino.

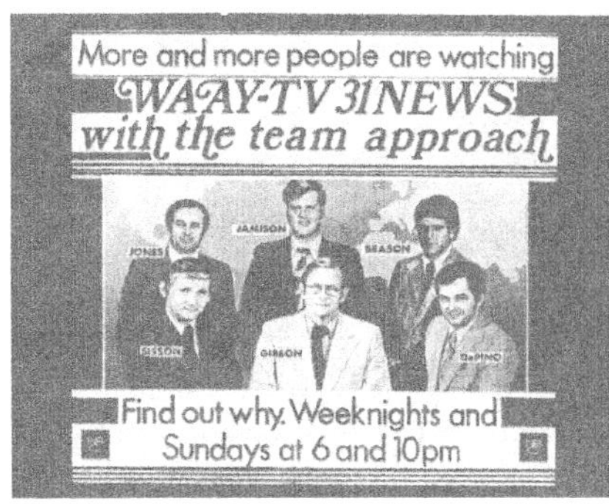

September - The anchor team grows by one more. Tony Beason a very credible reporter and anchor begin as seen in this photo We are on our way to Number ONE.

1974

February - We became Number One in news at 6 & 10 after three years working with the Consulting Firm of McHugh-Hoffman. We had not been #1 in news for over 6 years. We found out what it took overall to win in the ratings. Some of those things are still a secret lesson. A lesson M.D. Smith, IV never forgot.

The anchor team that was #1 in 1974 was (Left to Right standing) John Bradshaw (new-weather), Tony Beason (News), Dan Jamison (Sports). (Left to Right seated) Ed Sisson (News) Adrian Gibson (Anchor) and Sam Depino (Anchor).

April 3 & 4, 1974 - The April 3rd tornados (some people refer to April 4th, the day all the destruction was visible) gave us much publicity as the ONLY station with Radar until ALL TV stations were knocked off the air at 11:05 p.m. when the last tornado crossed the top of the mountain at Monte Sano and Panorama. Fifteen people in Madison County were killed. Many more in other North Alabama Counties. The stations were off the air for 2 days. Phil McHugh of McHugh-Hoffman Consulting was in town for a consulting visit at the time. Parkway City

Shopping Center was nearly destroyed. It was later rebuilt as a mall with covered center walkway. (Note: in 2001-2002 it was rebuilt again as "Parkway Place," a two-story and greatly enlarged mall.)

Governor George Wallace (then in a wheel chair) sent our station a personal letter (to News Director Adrian Gibson) thanking us for our Radar Coverage warning the people of North Alabama and likely saving many lives in the area.

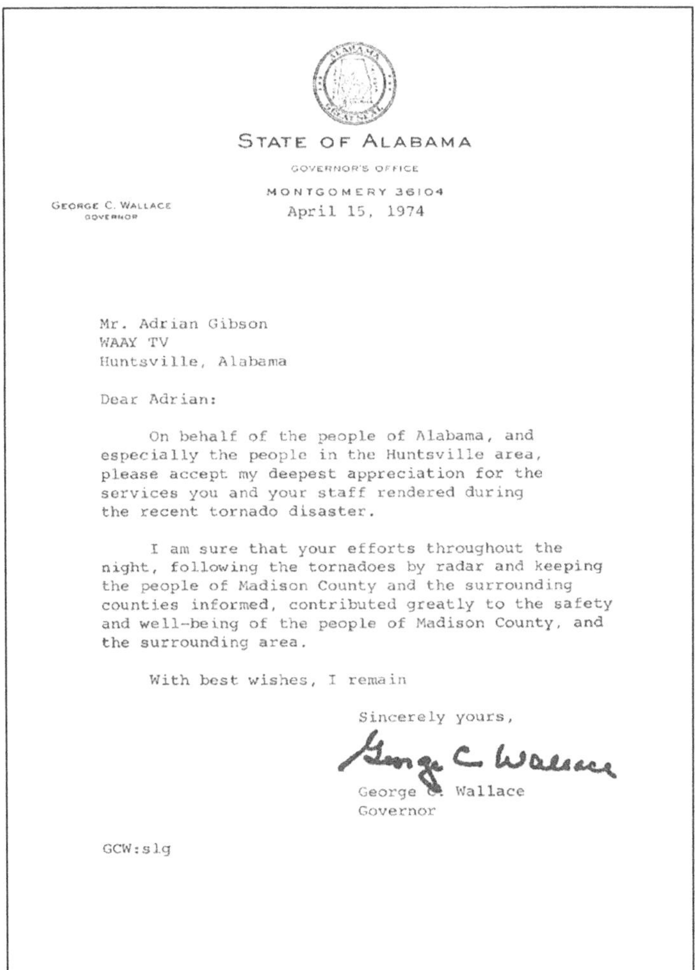

STATE OF ALABAMA

GOVERNOR'S OFFICE

MONTGOMERY 36104

GEORGE C. WALLACE
GOVERNOR

April 15, 1974

Mr. Adrian Gibson
WAAY TV
Huntsville, Alabama

Dear Adrian:

On behalf of the people of Alabama, and especially the people in the Huntsville area, please accept my deepest appreciation for the services you and your staff rendered during the recent tornado disaster.

I am sure that your efforts throughout the night, following the tornadoes by radar and keeping the people of Madison County and the surrounding counties informed, contributed greatly to the safety and well-being of the people of Madison County, and the surrounding area.

With best wishes, I remain

Sincerely yours,

George C. Wallace

George C. Wallace
Governor

GCW:slg

1975

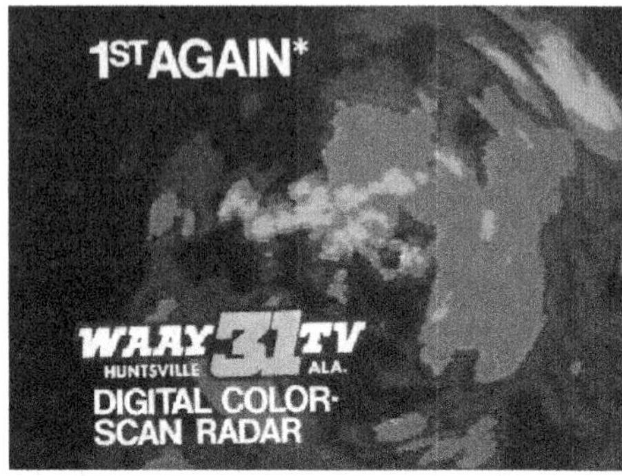

February - To improve our Weather Radar, we added digital color scan radar to it to make it easier for the viewers to see various intensities of storms. We were the first in Alabama with this innovation.

31 NEWS EVERYWHERE

April – We began what was to be a long running campaign of 31 News Everywhere with a bumper sticker campaign and cash & merchandise prizes. Viewers sent in photos of stickers all around the valley, including this one from Guntersville on the boat docks. Locations around the valley were filmed from a chopper. We put giant 31 Everywhere

letters on the ground. One seen here at Big Spring Park in Huntsville. Other locations were throughout the valley a as far as Florence.

We had a great set of music jingles used with on-air promos and radio spots. They basically said, “Look for us, we’ll be there, 31 News is Everywhere.” M.D. Smith, III and IV wrote the lyrics together for these jingles. The 60 second cuts could be split into two 30 second cuts for variety.

1976

April - New 1,000’ Tower - The board met and decided to replace our 400’ tower with a 1,000’ one. Took MDS IV, Cactus and MDS III meeting with Monte Sano Civic association and several homeowners in the area to get approval to build it. The erector set arrived on several flatbed trucks in late spring and on the next page you will see photos of the 6 months period it took to assemble each tower leg, brace and thousands of nuts and bolts that made up each 30’ section of tower. The job of clearing the land, pouring guy anchors, assembly, erection, adding transmission line, lights & power and finally placing the antenna on the top, was massive. It cost $300,000.00 to build. Here are photos of a newer tower built in 2004 but was the same height and thickness as the original “tall tower” we built that helped us secure the ABC network in 1977. It was the #1 network at the time.

It's the bi-centennial year (1776-1976) and before the July event, Channel 31 constructs a HUGE set, we call the '31 News Center' and the look and type style matches NBC's News Center. Left to Right are Ken Rainey (weather), Dave Stanley (news anchor), Adrian Gibson (news anchor) and Rick Davis (sports).

Two nice close-up shots of Dave Stanley with the TV Monitor and Two Way equipment. The other is of Ken Rainey on the freshly designed weather set.

"The Great American WAAY" was the Radio Theme during these few years and among other items, the "Great American WAAY" belt buckle in both pewter and bronze was created. Bob Baron was behind this idea and M.D. III hated to spend so much money. He got more of the pewter to give away, because they cost so much less. Here's a good close up of the WAAY buckle in Pewter. Also, here's the "1400 WNUE" (New) American Way in Bronze. These are rare collectors' items. Thanks to Don Roden for the last two images.

December - The Chamber of Commerce of Huntsville announced that due to declining interest and sponsor support, there would be no annual Christmas parade in Downtown Huntsville that year. Bob Baron with WAAY-Radio, M.D. Smith, III and M.D. Smith, IV all got together and decided that WAAY Radio and TV would put on an "All-Volunteer Christmas Parade." Everyone pitched in building their own floats, entries, and bands, and the parade went on. That was the FIRST and TWENTY years later, WAAY-TV still has the commitment to see that our all-volunteer Christmas Parade takes place. The location was moved from the route around the courthouse downtown, to a route starting beside the Huntsville Hilton and passing the Civic Center and looping the downtown area because of the size. The event attracted over 211 entries and close to 50,000 spectators.

December - Our 1,000-foot new tower nears completion (we began transmitting on it in January of '77) and we designed a special Christmas Card to send to all our clients and friends. Our tower is the tallest thing around and we figured it was so high, that even Santa might get tangled up trying to fly over it. Here is the outside and inside of the card.

1977

After 10 years as an NBC Affiliate, expanding into the Florence (Shoals) area, the competition from WOWL-TV became significant. With ABC we would have no competition in the quad cities area...AND ABC has just become the number ONE network in the nation. We have just finished building our 1,000-foot tower. We go back to ABC and make a presentation to change to our station which has become #1 in News ratings in the past few years (still a close race with channel 19) while WYUR is a distant third in the news ratings. ABC agrees we would be the best affiliate and agree to the change. Dan Whitsett and M.D. Smith, IV negotiated the switch in New York in June. The affiliate meeting is over, so Dan and M.D., IV fly to L.A. in July for the promotion manager's meeting. ABC affiliates are flying high with the ratings and are #1 for the second year. The theme for the fall campaign is "STILL THE ONE" using the popular hit song for a theme. The switch is made on December 11, 1977.

Here, my son Brent (age 9) is holding the full-page ad from the newspaper on that day.

A coffee mug was created with the "Still The One" logo and the 31 worked in. The reverse side of

the mug had the full size 31 logo and call letters. The heavy, serif "31" has been used for many years. Soon after we would adopt the more modern 31 logo with the circle around it similar to WABC-TV, Channel 7 in New York. (See 1979)

1978

January - M.D. Smith, III wrote a motivational letter to all the Radio personnel and I (MDS IV) got a copy and also delivered it to the 31 TV people because I thought it said such good things. You will find that complete letter with a signature of MDS III at the bottom, in another part of this book where individual stories reside. (Note: MDS III died in 1985 of cancer.)

February 1978 - Judy and M.D. Smith, IV fly to Las Vegas to pick up our new Live-31-Eye truck that we bought and had customized. We had to drive back in a winter storm that followed us all the way back on our trip. We saw the Grand Canyon covered in snow and no tourists there and only a few deer were around when we viewed the canyon view. From Little Rock to Memphis, it was all ice, and took eight hours at 5 MPH on the interstate most of the way, on a trip that should have taken two hours. This is a B&W photo of it just after the paint job was done on it. Here's the truck in summer of 1989 and at the 20th Anniversary of Man on the Moon.

1979

June - ABC network and Channel 31 continue to be tops in the ratings. ABC had used "We're The One" for a theme in 1978, but reverted to the music and logo of Still The One for the fall of 1979 and we make a new coffee cut sporting our new Circle 31 Logo for that year. Here is the back with the full station circle 31 logo. Notice the Red and Blue Racing Stripe that Scott Smith thought added a nice touch.

July - Bought our first Production Vehicle. It cost $2,000 used. It did the job and we didn't have to have production people use their own cars anymore and pay them mileage. Lew Koch was happy. You see the original front of the TV station (no dishes or radar), and the guy anchors for the new 1,000 tower are in front of the unit. (See 1981 for the next production truck we bought.)

1980

May - WAAY Radio installs and begins broadcasting from the new 50,000-watt

transmitter on Pulaski Pike for all daytime hours. At night, the station still used the downtown studio site for the 500-watt directional array. The current slogan for the hit music is This mug shows the 50,000 watts, Stereo, and "The Hot One."

May - Channel 31 has great ratings after several difficult years against Channel 19 and so we celebrate with a picnic on the front lawn. Some of those attending of interest: Bob Baron (Weatherman), Dan Whitsett (Sales Manager), Don Roden (was asst. chief engineer and after short time as engineer with Ch. 54 who came to town in 1985, went on to be Chief Engineer of channel 19 before leaving for private business) and Dick VanValkenburg, the News-Hound as he was called, who did all the wreck and hard news coverage with a Cannon Scoopic 16 film camera for years. He died from one of several heart attacks he had. Dick died in December of 1986.

The ratings for the book were a real reason to celebrate.

HOMES	Arbitron
6 p.m.	
31	71,000
19	59,000
48	14,000

# # # # # # # # #

10 p.m.	
31	51,000
19	39,000
48	12,000

December 11, 1980 - Eloise Smith Hanna died at age 87 in a nursing home.

1981

March - The FIRST "CHOPPER 31" takes to the air. This is the first time we have had a chopper for aerial photos (using fixed wing aircraft like in the past). It is one of many to serve Channel 31 since then.

July - Bought our second Production Vehicle. It was our first truck that cost $4,000 used. It replaced the Travelall that we traded in and got $1500 off the price. Lew Koch was again, happy. It was kept until May of '87 when it sold for $1,250 and we again stepped up to a newer van.

August 22 - (From 1979 to 1980, M.D. IV lost 100+ pounds with Weight Watchers and begins exercising in races, including the budding new sport, Triathlons.) There were none in Huntsville, so M.D. IV, with the help of Bettie Higgins, Channel 31 puts on our first of many annual 3-in-1 Triathlons. It's a 3-mile run, 6-mile bike and 400-meter swim at the Natatorium. We have 400 entrants (and more wanting to do it) the first year. It peaks at 600 entrants in 1985 and then drop back to 400 in the late 80's and early 90's) The Sub-Theme of this event has always been, "To Finish Is To Win."

1982

March 24 - Channel 48 has a fire in the early evening and the studio/transmitter building burns to the ground. They move to current site on the Parkway and rebuild a transmitter house at the old location.

April 13 - In record time, Channel 48 returns to the air. WAAY-TV runs a letter in the newspaper welcoming them back on the air.

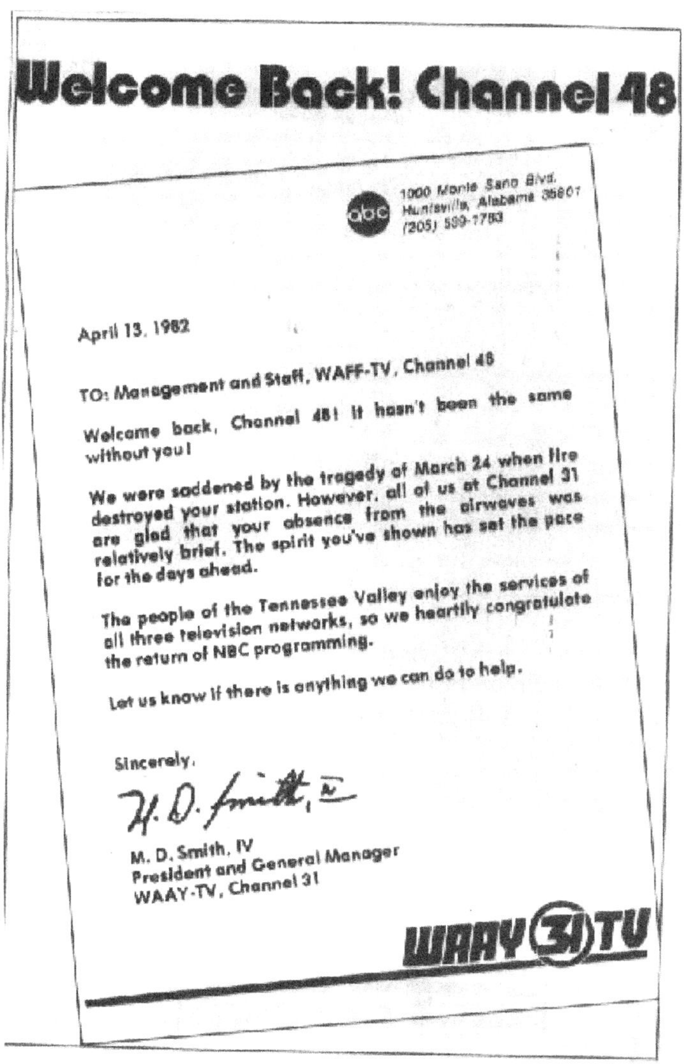

TUESDAY, APRIL 13, 1982 HUNTSVILLE NEWS Page 3

Welcome Back! Channel 48

abc
1000 Monte Sano Blvd.
Huntsville, Alabama 35801
(205) 539-1783

April 13, 1982

TO: Management and Staff, WAFF-TV, Channel 48

Welcome back, Channel 48! It hasn't been the same without you!

We were saddened by the tragedy of March 24 when fire destroyed your station. However, all of us at Channel 31 are glad that your absence from the airwaves was relatively brief. The spirit you've shown has set the pace for the days ahead.

The people of the Tennessee Valley enjoy the services of all three television networks, so we heartily congratulate the return of NBC programming.

Let us know if there is anything we can do to help.

Sincerely,

M. D. Smith, IV

M. D. Smith, IV
President and General Manager
WAAY-TV, Channel 31

WAAY 31 TV

1983

January 26 - Winds Of War - This was a huge reception and promotion put on by Channel 31 to promote ABC's Mini-Series, "The Winds of War" from the novel by Herman Wouk. It was an 18-hour TV production almost every night from February 6th to February 13th. We rented most of the Huntsville Space and Rocket Center (including the theatre) to put on this press party. Here's the invitation.

One of the highlights of the evening was a LIVE telephone hookup with Ralph Bellamy who played FDR in the series. He was asked questions of the press and gave answers that were quoted in local publications.

THE WINDS OF WAR

WAAY-TV and ABC TELEVISION
invite you to a private Press Preview of

HERMAN WOUK'S THE WINDS OF WAR

Wednesday, the twenty-sixth of January
Six O'clock P.M.
Alabama Space and Rocket Center

Step into four hours of the music, fashions and fads of 1939 that occupied America as the winds of war swirled in Europe. Experience a preview of what is acclaimed as television's most lavish and spectacular event.

Photo opportunities will be provided with very special guests.

Cocktails and Hors D'oeuvres

RSVP 533-3131 (Huntsville)
Bettie Higgins or Pat Spurrier

The press package face looked like this and went to all radio and newspapers in the area.

At the reception, members of the staff lined up to be introduced by General Manager "Commander" M.D. Smith. All women's uniforms were authentic. The women's came from the nation's only Army Corps Museum at Ft. McClellan. Note the hair styles.

Other photos of the event include:

1. Jim Marsh (anchor), Julie Martin (promotion), Carl Spurlock (producer) and Lee Baker (artist).

2. A Huntsville Times photographer admiring Heather Burns (anchor) and Helen Howard (reporter).

3. Here General Manager M.D. Smith Talks with the press just before the Ralph Bellamy's speaker-phone call.

4. Special guests also included REAL retired military officers and their wives.

5. This photo is of "Commander" M.D. Smith and Program Director Bettie Higgins who is wearing a 1942 WAC uniform and cap with the rank of Major.

6. In this typical ladies pose of 1942, we see Helen Howard, Heather Burns and Debi Benson.

7. A special cake with the Winds of War logo was created by Bettie Higgins.

8. This pose includes the tall Bob Baron (weathercaster) with Heather Burns (L) and Helen Howard and Debi Benson (R)

9. This photo includes News Director Mike Sullivan, Bettie Higgins and Mike's wife, Patty Sullivan (We drafted everybody.)

10. City Councilwoman Jane Mabry (tall-center) “joined up” with staff members mentioned above.

11. The 31 News Cars sported a triangular signboard promoting the February 6th series premier.

D-14 The Huntsville Times THURSDAY, FEBRUARY 3, 1983

‘Winds of War’

A Must-See TV Experience From

German Actor
Gunter Meisner
As Adolf Hitler

On Sunday, “The Winds of War” will begin to whirl across our TV screens. It is a story that edges in like a chill and gusty winter night, pauses occasionally for the spring breezes of life’s pleasures and gathers speed toward a fearsome tornado of war that will sweep up and forever change the fates of all its characters.

At times, the people involved in the production of this massive novel-for-TV movie must have felt in a war zone themselves. After years of struggling to gain the novel rights, the deal was finally struck between Paramount and author Herman Wouk on June 16, 1977. ABC, reportedly agreeing to pay $1 million per hour to air it, originally planned to show it during the 1979-80 season.

Changes in studio personnel, actors’ strikes, switches in directors, producers and major cast members — all delayed the film.

A Review By Judi Moon Times Extra Sections Editor

Eventually, Wouk, who had retained final approval of the movie adaptation, decided to write the screenplay himself after the man who had begun it died of a heart attack.

The cameras rolled at last on Dec. 1, 1980. Shooting, the longest ever for a TV project, ended in December 1981, four days ahead of the 206-day schedule. Miniature work then was completed

12. As a result of the WOW party, here is one newspaper article by Judi Moon from “The Huntsville Times.”

'The Winds of War'

A special preview of ABC-TV's "The Winds of War" was shown last night at the Space and Rocket Center, hosted by WAAY-TV, Ch. 31. Outfitted for the occasion in authentic uniforms borrowed from the WAC Museum in Fort Maclellan are, from left, Bettie Higgins, Heather Burns, Helen Howard and Debi Benson, all staffers at WAAY. (Staff Photo by Jim Taylor)

13. Another newspaper photo with our women "Saluting their superior officer."

This was a major event and promotion of it. The ABC Network gave us special credit for our local efforts in a future Affiliate Meeting the next June. M.D. Smith, III the CEO of Smith Broadcasting was particularly proud of this event since he loved promotion in Radio and TV stations.

****

November - WAAY Radio installs the New KAHN stereo system, which was incompatible with the other 2 systems on the market, with Motorola later becoming the system used in the receivers in cars. Install new stereo console, tape recorders and had a major ribbon cutting with Joe Davis, Mayor of Huntsville cutting the ribbons on the new stereo facility. (Glen Buxton was manager of WAAY radio at this time, Wayne Johnson, previous manager having been transferred to TV in sales support work.)

1985

February - It's one of the city's worst ice storms that lasted 5 days. Power out to thousands of homes in the area. Takes many days to get it all restored. Just after the 5th day of the ice storm, Mr. Smith, III suffered a minor heart attack. He never seemed to recover well and continued to have chest problems and trouble breathing. In the summer of 1985 it was discovered he had lung cancer in a very advanced stage. The board decided with the declining ratings of Top 40 and the health of Mr. Smith, III that the AM stations would go with an Adult Contemporary Format from Satellite Music Corporation. WAAY changed formats in August and WNUE followed in September.

M.D. Smith, III died from lung cancer on September 20, 1985.

Kirby Jones Smith (She preferred to be called Elizabeth Kirby Smith) becomes Chairman of the Board, with members M.D. Smith, IV, President —Anita Smith Johnson, Vice President—Ralph Dowdy, Member.

April - Channel 54 TV signs on the air. They are an independent station with no network affiliation. They run movies and syndicated programs. Later, they become affiliated with the FOX television network, as it starts from only a few hours a week to programming every night of the week.

May – Another logo coffee mug was this one containing our News slogan, "The Clear Choice" on the side of it.

1986

July - Remembering the Ice Storm and Tornadoes of the past, Channel 31 decided to install a monster Caterpillar Diesel Generator to power the entire station when the power goes out in the future. It was one of the things M.D. Smith, III talked about when he was alive and always wanted to do. We are the only station to do this (until AFTER the 1989 Tornadoes).

1987

M.D. Smith, V (Dee) goes to work in news as a weekend photographer. In 1989 he transfers to the Engineering Department in light maintenance and purchase and inventory control.

July - Scott Smith joins TV as Art Director after several years working for printing companies as commercial artist.

November 30 - First 5:00 p.m. newscast begins on Ch. 31. We were the first in the market and for 2 years had the ONLY 5:00 show. It became #1 over a sit-com and a game show and finally Ch. 19 added their own version that had less than half the numbers our show did by then. Ch. 48 expanded their 5:30 news back to 5:00 in 1992.

November - The Fall 1987 Ratings in Both Nielsen and Arbitron show 31 a big leader.

Arbitron

5:30 p.m.

Station	Rating	Share	Households
19	13	22	41,000
31	11	19	33,000
48	18	32	58,000
54	4	8	14,000

6 O'Clock

Station	Rating	Share	Households
19	14	21	44,000
31	20	32	63,000
48	13	21	45,000
54	5	8	17,000

10 O'Clock

Station	Rating	Share	Households
19	10	23	33,000
31	16	37	47,000
48	9	21	31,000
54	2	4	5,000

Prime Time
(7-10 p.m. Mon-Sat and 6-10 Sun)

Station	Rating	Share	Households
19	14	23	48,000
31	14	23	43,000
48	14	23	44,000
54	3	5	10,000

Sign-on to Sign-off

Station	Rating	Share	Households
19	8	24	26,000
31	7	21	21,000
48	7	21	22,000
54	2	7	7,000

Nielsen

5:30 p.m.

Station	Rating	Share	Households
19	15	26	45,000
31	10	18	31,000
48	14	25	42,000
54	5	9	17,000

6 O'Clock

Station	Rating	Share	Households
19	17	26	50,000
31	19	29	58,000
48	11	17	33,000
54	4	7	16,000

10 O'Clock

Station	Rating	Share	Households
19	13	29	42,000
31	14	30	43,000
48	8	18	27,000
54	3	6	9,000

Prime Time
(7-10 p.m. Mon-Sat and 6-10 Sun)

Station	Rating	Share	Households
19	16	26	52,000
31	13	20	41,000
48	15	23	45,000
54	3	5	11,000

Sign-on to Sign-off
(7 a.m. to 1 a.m. Sun-Sat)

Station	Rating	Share	Households
19	9	27	29,000
31	7	19	21,000
48	7	20	21,000
54	2	6	7,000

Nielsen Metro

(Madison, Morgan, Limestone)

5:30 p.m.

Station	Rating	Share
19	15	30
31	10	20
48	15	29
54	4	9

6 O'Clock

Station	Rating	Share
19	18	31
31	18	31
48	13	22
54	4	7

10 O'Clock

Station	Rating	Share
19	14	29
31	13	28
48	10	22
54	4	7

Prime Time
(7-10 p.m. Mon-Sat and 6-10 Sun)

Station	Rating	Share
19	17	28
31	13	22
48	17	28
54	3	5

Sign-on to Sign-off
(7 a.m. to 1 a.m. Sun-Sat)

Station	Rating	Share
19	9	29
31	7	21
48	8	23
54	2	7

(No household figures are available for Nielsen metro survey. All numbers are copyrighted by Arbitron and Nielsen from their November 1987 surveys.)

December - Brent Smith begins work striping La-Kart carts for production. Later transfers and becomes a switcher and master control operator.

1988

February 13-28 Calgary, Canada ABC covered many Olympic Games. Here's a special cup that had a gold leaf. Circle 31 logo on one side of it and the Logo for the Calgary 1988 Winter Games on the reverse side. We ordered these special made from ABC and they were quite expensive and a treasured client gift.

June - Losses continued at WAAY radio in spite of the reduced payroll from the satellite operation and it was decided to take it off the air in the summer of 1988 while we looked for a buyer for the station. No buyer was found immediately.

July - The board decides to give one last major try at WNUE to make it successful and drop the AC and satellite service and go back to live DJ's and the HIT (Top 40) radio sounds that made the station famous. It was "NUE HITS 14" and was a great sound. The logo was red letters with a yellow outline that looked like fire, over a blue splash of water underneath. Looks great in color. The new logo designed by Scott Smith was put on a new coffee cup fire red over yellow, "Nue Hits 14, WNUE." It was good old TOP 40 in 1988 and music was still on 45 RPM records but was beginning a transition to CD's. Ratings did jump up several notches and the station was well listened to in the Ft. Walton area. However, getting and/or keeping competent sales personnel was all but impossible and after one year and two months (Sept. 1989), with losses large and continuing to grow, the station was also taken off the air and put on the market to sell.

September - A frequency is opened in South Pittsburg, Tennessee (near Chattanooga) and M.D. Smith and Robert "Cactus" Gay apply for it in the name of Smith Broadcasting. Later Cactus becomes the other partner. The Construction Permit (CP) was authorized in October 1989, but the change in Authorization to Marson Broadcasting (the new partnership) did not take place until April 23, 1990. Construction began immediately.

1989

May - Sam Floyd of United Broadcasting buys WAAY radio for $300,000 and begins broadcasting in the early summer of 1989 with a Gospel Religious format. Unfortunately, with TV Evangelist Jim Bakker going to jail and the other TV ministries scandals, religious broadcasting took a big drop in support.

1989

Kimberly Johnson (daughter of Anita and Wayne) comes to work at Ch. 31 after working previously at WAAY Radio. She is in TV traffic department.

November 15 - A freak tornado rips through Huntsville and destroys Airport Road businesses, apartments, Jones Valley Elementary School and many homes on Tony Drive. Eighteen people were killed. Channel 31 was the ONLY station to have a generator and be able to stay on the air that night for extensive live and taped coverage. We got much praise from that wise move. Sometime later, other stations got a generator, too. Later the TV station raised $26,000.00 for the Red Cross Disaster funds (see Feb. '90 for details).

1990

February 12 Richard Matthews of the American Red Cross receives a check from M.D. Smith, IV for $26,000.00. This is the proceeds from sale of tapes of our news coverage of the Nov. '89 tornado. You can see the "large" check

presented to the Red Cross. The current news set of 31 News can be seen in the background with the backlighted "sky" and leaves in the foreground.

March - The STAR 31 Logo is born. Scott Smith was asked to rework the logo without the circle around it and while doing so, he tested putting a STAR in the 31. It was approved and, in a few months, it was on the building, cars, coffee cups and certainly on the air.

Summer - No buyer found for WNUE, which has been off the air since fall of 1989. Some equipment is used for the new radio station, WKXJ-FM, a Class C2 station being built in So. Pittsburg/Chattanooga, Tennessee on 97.3 MHz.

November - WKXJ-FM goes on the air on November 5, 1990. This is the first bumper sticker for 97-Kicks. It is a "Hot Adult Contemporary" format. M.D. Smith, IV and Robert A. (Cactus) Gay own the station under Marson Broadcasting. (Marson is an old name used by M.D. Smith, III during the war years in his manufacturing plant, Marson Company. It was the "Mar" from Marcellus and the "Son" from Davidson that formed the name.) The "oldies" music was bought on 10" Reel-to-Reel tapes and dubbed with great effort to the old "carts"(many came from WAAY and WNUE radio). Carts were the cartridge tapes that played commercials and music at the radio stations in the 80's to keep music sounding better and not getting scratched and dirty as in the early days when the actual 45 record was played on the air. The looked very much like the old 8-track music tapes but were built differently. However, all the NEW music of the 90's came on CD and was played on a home type CD player. Eventually a new music "oldies" library was bought and the cart tapes were retired for music.

A few years later, all commercials were put on computer disk and the cart machines were taken out of service forever.

September - The general manager, Harmon (Dutch) VanBuskirk, of WNUE for 25 years, finds a partner, Neil Braswell and decide to buy just the frequency and transmitter & tower and lease the building for 2 years. Dutch gets back on the air in October 1990 with a "Talk" format. It is satellite news and talk and Dutch has some live phone-in talk shows.

1991

May - After losing money for almost two years at the newly named WAAJ, (the old WAAY-Radio on 1550) the station goes off the air again and comes back to Smith Broadcasting.

November - Channel-31-TV ratings come in to show our station the dominate news leader in Huntsville. The station is #1 for ALL newscasts, seven days a week in ratings, share and households. This is the strongest position the TV station has ever been in. The other stations continue to cut staff. Now both channel 48 and channel 19 have staffs of about 75, while channel 31 has a staff of over 125 with the plans to begin a new noon newscast on June 1, 1992.

1992

February - No buyers were found for WAAY-AM, so the equipment was advertised to be sold. In February the board voted to demolish the WAAY studios building (on Church Street near downtown) and develop the property in some other way. The building is leveled in 45 minutes with a bulldozer on Thursday, March 12, 1992. Transmitter sold to Bishop Willis for his Memphis 50kw station. Later a minority partnership bought the license, towers, and daytime property on Pulaski Pike and put WAAJ back on the air with the old 5,000-watt transmitter and begin a religious-gospel format.

(End of original '55 - '92 M.D.S. IV Update section)

1992

June 1 - Ch. 31 began NOON newscasts Monday through Friday.

June 8 - Ch. 31 becomes the first station to Close Caption ALL its local newscast for the hearing impaired. Ch. 19 began a year later. There was a large reception and celebration held. The local AIDB (Alabama Institute for the Deaf and Blind) and other officials and dignitaries attended the announcement at the Huntsville Hilton.

Here, Mayor Steve Hettinger addresses the audience (with person "signing" what is being said). There was a special cake both the WAAY-31-TV logo and the [CC] logo on it to honor the event. This also happened to be Mr. and Mrs. M.D. Smith, IV's thirty-first (31) anniversary that same night. Judy Smith received a special flower arrangement and a "Circle 31" gold necklace made the same as the 31 logo pin anchors wore a few years before.

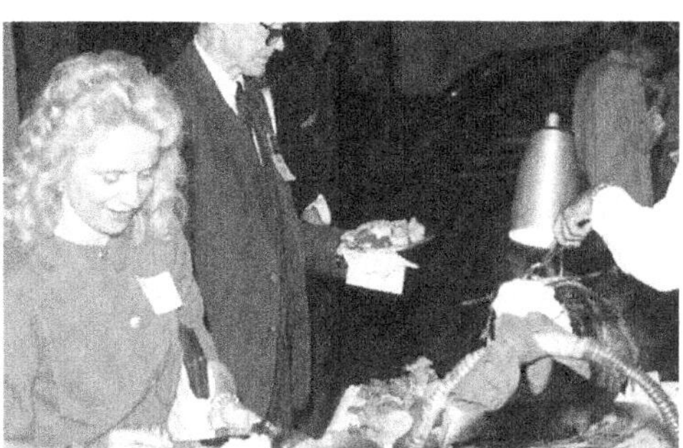

August 22 - The 12th Annual 3-in-1 Triathlon takes place. A special Star 31 Mug was created for winners in the age divisions. As always, the slogan "To Finish Is To Win" was on it.

1993

January 3 - Sunday, 8 a.m. newscast begins on Channel 31, following ABC's beginning of Good Morning America, Sunday edition from 7-8 a.m.

April 15 - It is announced that Channel 31 has re-hired Bob Baron who left Channel 48 and he will be doing weather forecasts along with Gary Dobbs after a year of working behind the scenes (due to previous contract restrictions keeping him off the air for 12 months.) Several nights later, severe weather hits the area and the new "31 Storm Tracker" does a superb job tracking and predicting where storms will move. It is the only tracker on TV in the valley. Others add similar capabilities some months later.

April 17-19 - Ch. 31 begins weekend early morning news from 6:00 to 7:00 a.m. on Saturday and Sunday.

Ch. 31's Monday through Friday 6:30-7:00 a.m. news expands to a full Hour from 6:00 to 7:00 a.m. and we move the WAAY-TOO-EARLY program back to 5:30 a.m. (W2E later expands to one hour also beginning at 5 a.m.)

Summer - We decide to become a 24-Hour News Station presenting MORE NEWS around the clock. Our new coffee cup logo for this year. We are already doing news-briefs every hour on the hour, 24 hours a day, and 7 days a week. Sports was included in the 24-Hour on the hourly Newsbreaks.

August 21 - The 13th annual "31 3-in-1 Triathlon" takes place at Old airport and Natatorium.

31 SPORTS
24 HOURS

November - WKXJ-FM, 97Kicks begins broadcasting with a 99.3 Translator in the Chattanooga Metro area. Now with the 50,000-watt main transmitter on 97.3 covering the rural area and the new translator covering Chattanooga, the future looks brighter for the profit potential of the radio station. On this, its 3rd anniversary (Nov. 5) it still is not making significant profit but paying its own way. We decide to make the sound even "hotter" and are successful in getting teens and young adults with the "Hot Jammin' Kicks-FM" sound and logo (Both this and original sticker artwork done by Scott Smith).

1994

April 15 - Bob Baron Joins Gary Dobbs on air...we now have the top one and two weather people in the valley on our station.

1995

March 1 – HiWAAY Information Service becomes a reality. We decide to become our own Internet Access Provider for our TV station, and supply to the public at the same time. Bryan Smith comes to work as Manager to this fledgling operation. We premiered at the IEEE Computer show in February and sold subscriptions even though the HiWAAY was not paved. We do get operational on the first, but there were quite a few start-up problems. These problems are ironed out in the first several months.

We hoped to have 200 subscribers after 12 months. In less than a year, we had over 2,000. Buying NUANCE Internet Services in August of 1995 added about 250 subscribers to the subscriber base. Soon we added a slogan, "Real Internet—Real Fast" to the original globe logo we created in late 1994.

1995

July 7 – LIVESTAR-31 arrives in town. This is Channel 31's first Satellite Uplink Truck. It is a monster and bigger than the one Channel 19 has purchased that comes to town a week later. Now we can do LIVE broadcast from anywhere in our viewing area, which previously were blocked for microwave by the mountains.

August 19 - The 14th annual "31 3-in-1 Triathlon" takes place at Old airport and Natatorium.

August 25 - SHOALS STAR RADAR comes to the valley. This is a high-powered Enterprise Electronics Corporation (EEC) weather radar mounted in a Winnebago Motor Home. It is taken to the Shoals and put in service as an early warning radar for areas west to the Mississippi river. It has the Baron processor on it, which can see 16 levels of rain.

November - Channel 31 announces the opening of THREE new area bureaus. They are located in Guntersville (Marshall County), Scottsboro (Jackson/DeKalb counties) and Fayetteville, TN (for Tennessee counties). This brings to FIVE, the area bureaus covering the whole 31 viewing area with regional/area bureaus for a local focus

on news, un-matched by any other TV station. We also can get to news events faster with AIRSTAR, a twin-turbine prop-jet MU-s aircraft.

1996

January 1 - Beginning announcements of "**31 DOPPLER MAX"** on Channel 31. We are testing out the most sensitive Doppler radar in the valley. The antenna is located on top of a specially constructed, 120-foot, 3 section base tower, supporting the massive EIGHTEEN FOOT RADOME housing the 14' diameter dish antenna. It is the best technology for Doppler weather radar in the Southeast. It has the same transmitter power as the U.S. Weather Service at the airport, but has a larger dish for more gain, and the state-of-the-art Baron Processor with 16 levels of color intensity for rain, and special Doppler functions not available with any other radar. The premier of "31 DOPPLER MAX" is on January 17, 1996.

As a part of this Doppler Max "kick-off" over 100,000 bumper stickers where distributed to announce the most sensitive radar in the valley.

December 7 - The tradition continues - WAAY-31-TV held its 20th Annual Christmas Parade in Downtown Huntsville. In order to commemorate this tradition that started in December of 1976, a special Anniversary Patch was made.

1997

March 1 HiWAAY has TWO-YEAR Anniversary. It has grown to over 5,000 individual subscribers and a large number of corporate network and web-site customers. The offices and computer equipment were relocated to spacious accommodations at Market Square on Memorial Parkway.

May 23 - Mrs. M.D. (Elizabeth Kirby Jones) Smith, III (age 80) passed away after a long period of declining health. M.D. Smith, IV continues as President, CEO and General Manager of WAAY-TV. Anita Smith (sister) and M.D. Smith, IV now co-own Smith Broadcasting, Inc. and the TV station.

June - The Storm Team (weather team) of Gary Dobbs, Adrian Gibson and Kristen Cornett take to the air. Kristen is also trained to drive and operate a remote weather station (Feb. 98) with GPS tracking called the Storm Chaser, Mobile Weather Center. There is much focus on the weather in these years.

1998

December - HiWAAY Information Services, the largest ISP in Huntsville, buys all of AirNet and the on-line business of Traveler, Inc. to become the largest independent ISP in Alabama. Now with over 20,000 dial up customers and many hundreds of commercial customers in North Alabama. There are also POP's (Point of Presence) in Birmingham and Montgomery. Under the leadership of Mark Derrick, HiWAAY continues to grow and prosper.

1999

January - M.D. Smith, IV elects to move from Channel 31 on Monte Sano, where he has had offices for 36 years, to the home office location on Clinton Street in the Civic Plaza building. Because of the merger of HiWAAY and the other ISP's, this central location makes it easier to direct attention to both businesses. Redge Swing is promoted to General Manager of WAAY-TV while M.D. Smith, IV remains President and CEO of Smith Broadcasting, the parent corporation. Dan

Whitsett is in charge of corporate sales but concentrating on work with HiWAAY.

March 1 - HiWAAY Information Services has FOUR-YEAR Anniversary. The offices at Market Square on Memorial Parkway had

to be greatly expanded and currently a build-out across the mall in a vacant area is underway to support the nearly doubling of staff after the acquisition of AirNet and Traveler. A special Blue & Gold logo coffee cup (collector's item) was produced for clients.

May 18 - It is announced to the staff that WAAY-TV is being sold to Grapevine Communications of Atlanta. The main reason given is to concentrate the family business efforts into the expanding Internet business and use the needed capital to grow at a rate equal to other Internet Service Providers (ISP's) in Alabama. The sale had to wait for FCC approval. It was granted in July and the final order was received in early September. WAAY-TV, Channel 31 officially becomes owned and operated by Grapevine/GoCom Communications on October 1, 1999. Smith Broadcasting continues in the Internet Communications business, the "broadcasting" medium for the 21st Century.

The "WAAY TOO EARLY" morning show was cancelled with the sale and M.D. Smith, IV was presented with the logo that was on the set for many years signed by every member of the morning crew. It was featured in a newspaper article about selling the station that was called "WAAY To Go."

WAAY to go

Broadcasting is a family business for M.D. Smith IV, left, and his father, M.D. Smith III, in a 1983 photo taken when WAAY radio went to 50,000 watts.

With the sale of his TV station, M.D. Smith signals the end of an era in local broadcasting

Bryan Bacon/Huntsville Times

M.D. Smith's 36-year broadcasting career came to a close on Oct. 1 when Grapevine Communications acquired WAAY-TV from Smith for $49 million.

2000

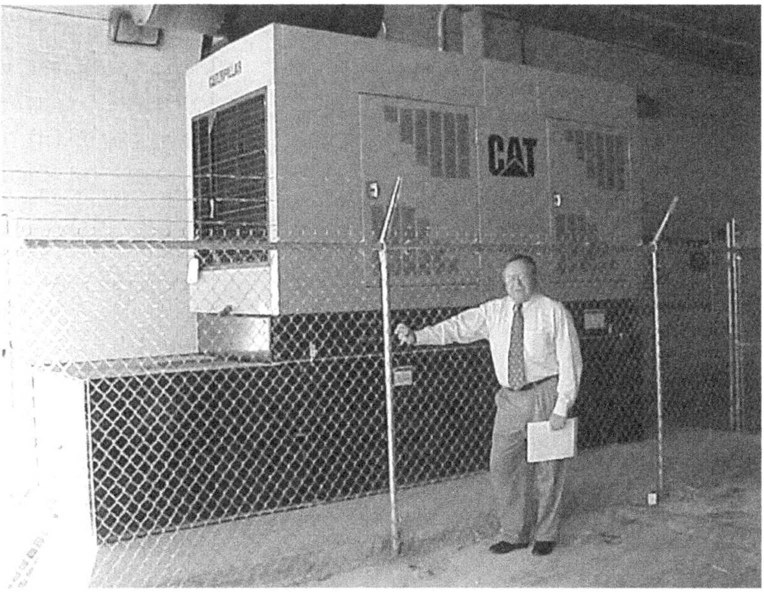

January 20 - HiWAAY Information Services continues to purchase other ISP's in Alabama. One of the largest is Idea-Works of Birmingham adding over 3,000 dial up subscribers and several hundred commercial customers. Redge Swing, who was in sales at Channel 31 for 30 years and joined HiWAAY in December of 1999, is promoted to CAO (Chief Administrative Officer) of HiWAAY. Here he stands by the 200,000-Watt Caterpillar Generator that powers the battery backup systems in case of a power failure.

HiWAAY
INTERNET SERVICES
www.hiwaay.net

March 7 - HiWAAY Internet (formerly "Information") Services has FIVE-YEAR Anniversary. Having bought other ISP's in Huntsville and Birmingham, the HiWAAY has grown to 22,000 dial-up scribers and over 2,000 commercial customers. HiWAAY will continue to grow larger with major expansion of DSL (Digital Subscriber Line) and ADSL services beginning this month.

March 20 - Ground is broken on the old WAAY Radio property for a new 480-unit self-storage facility to be called, Smith Store It. Located

SmithStoreit SELF-STORAGE

at 1015 County Road, this is the 5-acre site of WAAY Radio before it closed in the 80's. The old radio station on 1550 khz is still on the air, but broadcasting from the daytime site (see May 1980) north of town. This is a re-birth of a Smith property bought in the 50's. The web page for it is smithstoreit.com and completion is expected by fall 2000.

August 7 - Redge Swing is promoted to CAO (Chief Administrative Officer) AND General Manager of HiWAAY. HiWAAY is on track for growth and expansion in the state of Alabama, as a major Birmingham area marketing campaign begins. Construction of a new "Data Center" will begin in Birmingham in January of 2001.

August 19 - 20TH ANNUAL 3-in-1 TRIATHLON is conducted at Old Airport. Smith Broadcasting is the sponsor and has been for 20 years in a row.

September - Even though Channel 31 is no longer owned by Smith Broadcasting, they are still using the Star 31 logo and have added the web site address to this gold rimmed cup.

2001

January 2 "Smith Store It" opens for business. Grand opening with ribbon cutting by Mayor of Huntsville, Loretta Spencer, takes place May 12, 2001. Brent Smith and Mike Click (who married Allison Smith) have been the two people working full time with the storage operation since construction began in March of 2000. Scott Smith designed the logo for SSI and is doing other graphic work for signs and printed materials.

M.D. Smith, V ("Dee") had been involved with the web site design and updates. Dee and his wife Kristen had their first child on 01-01-01 (it was snowing this day) and named him M.D. Smith, VI and call him Marc. (www.smithstoreit.com)

August 18 - 21ST ANNUAL 3-in-1 TRIATHLON is conducted at Old Airport. HiWAAY Internet Services is the sponsor and this is the 21st year of Smith's Sponsorship of the event having put on the first one in 1981.

2002

January 11 - Martin Smith branches out from the family business when he purchases Auto Accessory Express at 202 Ward Avenue, Huntsville. He is the primary owner of the partnership with M.D. Smith, IV. Here is a photo of Martin and M.D. Smith outside shop. Also, here's the re-designed logo as well as the original above.

AUTO
ACCESSORY
EXPRESS

April 12 - M.D. Smith, IV was honored at his Alma Mater, The University of Alabama, with the "Betsy Plank Distinguished Achievement" award. Here Cully (E. Culpepper) Clark, Dean of the College of Communication and Information Sciences, presents the award at a ceremony in the Old Union Building. The refurbished building is once again the home of the Communications School as it was when Smith attended and graduated in 1963.

August – SmithStoreIt.com Self Storage expands with purchase of Highway 72 East Mini Storage. In September, a new billboard went up on the Highway and soon other improvements will be made to this existing facility. While smaller than SSI on County Road, it's in a very good location with high traffic count.

2003

March 5 - HiWAAY Internet Services 8th birthday. Here is a TV commercial still photo of the falling dominoes, which depicted other ISP's falling by the wayside, while HiWAAY remained rock solid. In a later commercial, HiWAAY is shown to have all the "building blocks" needed for a complete internet (castle) solution to business and personal desires.

Certainly, it has been more difficult for ISP's in the last few years and a lot of them have folded up. HiWAAY continues to improve the bottom line and while not making much money, we are not losing it, either. On this occasion, the 8th birthday, we had a cake to share with staff and friends. Gathered behind the cake are (L to R) M.D. Smith, Redge Swing (GM), Charles Boley (dept. manager and one of the oldest employees) and Scott Swing And closer, M.D. Smith, IV cuts the 8th anniversary cake to start the celebration.

August 23, 2003 - The 31 Alumni Reunion

A gathering of as many staff members as we could locate from the years 1963 to 1999. MDS IV created the "31 Alumni.com" website before the event. Here will be found tons of stories of past 31 Alumni (and current employees) along with many photos of them then as well as now. Great addition to the history you are reading now. Here is the opening paragraph of my "welcome" speech from the web site.

"To have 50% of the adult audience for a newscast is powerful stuff. Yet, that's what Channel 31 did in the 70's, in the 80's and early 90's. We hit numbers that made history. All of you here tonight were a part of either the building process to get there or being on top and trying to stay there. Both ways, it was always a challenge, usually fun and at the same time we always had to make a profit to stay in business and grow."

The photo is of Ken Rainey and Dave Stanley (weather and news anchors of mid-70's) with me. **You will find many photos of the reunion taken at Ditto on that date and news about 31 that has occurred since, like the tower collapse at the website, if it is still in existence. Visit http://www.31alumni.com/ and the companion http://www.31alumni.com/31histor.htm which is this history page you have been reading. Again, if it still exists on the Internet. This book contains far more information and photos than the web site does, but most photos on the web are in color.**

2005

March – HiWAAY Internet Services is TEN years old this month. It has been ten amazing years to see the growth of a small branch of WAAY-TV, which currently has almost the same number of employees as channel 31 does today. Two new products have been added this year to compliment the range of services that HiWAAY offers.

The first one is the Rocket Access accelerator for dial-up access. Running this software will increase speeds up to five times normal. Of course, if you want blazing speed, you sign up for Broadband DSL service.

The other service is the HiWAAY Computer Clinic for sick computers. The clinic will wipe all the viruses, spyware, worms, Trojans and other bad infections from your computer and add several protections programs to keep your computer healthy and non-infected in the future.

June 2005 - Smith Store It self-storage expands our 72 East location with thirty-eight CLIMATE Control and 82 standard units for a total of 400 spaces. Brent Smith becomes General Manager of both locations. **https://smithstoreit.com/**

2006

May – HiWAAY Internet Services launches the **Network Services Division.** Customers can now have HiWAAY install and maintain their LANS and equipment in addition to all their hosting, serving, and connection needs to the internet.

July – HiWAAY Internet Services moves to a new and more modern facility on Drake Avenue. More spacious parking in a village park atmosphere filled with other technical companies and doctor's offices.

2009

June – After M.D. Smith, IV takes over as General Manager of HiWAAY in late 2008, son Bryan Smith moves from the Birmingham office to Huntsville to become Operations Manager and a more key part of the family business.

2010

March 2010 - HiWAAY Internet Services Celebrates our 15th Anniversary.

May 1, 2011 - HiWAAY Internet Services was bought by Jason Mohprasit (employee of Hiwaay) & partner. The Smith corporation had not been successful at making a profit, due to the extremely large staff and loss of customers to the now giants in the business, like cable internet companies and telephone companies competing at far lower prices than we could offer.

2018

October 4, 2018 - As this printed version is being written, SmithStoreit.com continues to operate the two locations of the Self-Storage facilities, one at 1015 County Road between Cook Ave. and Church Street and the other location at 2386 Highway 72 East a mile past Shields road heading towards Scottsboro.

Son Martin Smith is the Resident Manager of the 72-East facility and Jeri Lamon is the Resident Manager of County Road.

Brent Smith is the General Manager of both facilities and reports directly to M.D. Smith, IV.

M.D. still maintains an office in downtown Huntsville with Sadie Rice, business manager and corporate accounting supervisor of forty-four years. Her first day at work was May 30, 1974, Memorial Day. (Yes, you heard that correctly. It was not a holiday in those years.)

Part 2

Running the TV Station in 1963-1965

By M.D. Smith, IV

I had graduated from the University of Alabama, majoring in Radio & TV, in 1963. I worked for WAAY AM Radio for some months while negotiating to buy the lone TV station in Huntsville, Alabama. We had planned to build a new station on Channel 25, along with another group with a license to build another station on Channel 19. The owners of Channel 31 were worried about two high-power new TV stations coming on the air shortly, approached my father, and a sale completed in June of that year. The FCC approved it, and we took over in November of 1963, two weeks before the assassination of President Kennedy. There were no commercials run for five days which hurt revenue after our taking over. Some sponsors had already dropped off because of the change in ownership.

I was the Operations Manager, working in tandem with the Sales Manager, Maury Farrell in those years. My father was technically General Manager, but spent little time at the station, working from his home office and running the radio stations which were the big money makers at the time. Our radio station grossed more money that the TV station did for the first year and "netted" more money for quite a few more years.

When we took over, only 15 people worked for the station. Some were just part-time or did both sales work during the day, and the news, sports or weather in the evening. We soon hired two newsmen (replacing two other people) and two photographers for our news team to take over for the single male news anchor, who left shortly for another job, fearing he might get fired.

That made a staff of 17 total people running the TV station with a LIVE kid's show every afternoon featuring personality Johnny Evans (Osborne) who was a salesman during other times of the day, and a very good one I might add. It was a real "door-opener" to be a TV personality when you called on potential advertisers. With Maury Farrell and Johnny Evans, we usually had one other full-time person selling for the station.

Almost everyone had multiple jobs to perform each day, including me. I came up through the production, promotion, and audience generating part of the broadcasting business and had some engineering background getting my FCC First Phone engineer's license in 1960. I was working night shifts at radio stations in Tuscaloosa while in school, and summers at one of our stations in Ft. Walton or Huntsville doing the midnight to dawn shifts which required an FCC First Phone on duty and announcers with better voices moved up to shifts with higher audiences during the vacation times. So, I had done a lot of Radio announcing & DJ work, but not very much in "Prime Drive-Times."

I took a college course in Electrical Engineering and became an Amateur (Ham) Radio licensee and operator in 1962. I built a lot of my equipment in those years, some in kits and some from scratch with just a schematic and scrounged parts. I did know how things worked and how to fix a lot of the old tube equipment.

Thus, in addition to managing the staff, outside of sales, which was engineering, news, production, commercial writing, traffic (putting programs & commercials on the daily log & timing segments), film man who handled slides and 16mm film commercials and programs), the single artist who painted B&W and shades of gray cards that were then photographed and made into B&W slides for commercials, and a part-time janitor. I did the hiring and firing of everyone but sales, and as we became more successful, we added a few to the various departments.

We were soon doing a morning 9:00 a.m. thirty-minute show called "Coffee Break." In addition to staff people, like the film man and an engineer running the two cameras, Maury Farrell and several different staff members hosted the show. It was usually the traffic manager who had good looks, and she could fill in the comments since Maury had been a live sports announcer in earlier years and could have done it all if he needed to. But it was a nice touch with two people. We needed a director to switch between the two cameras, line up shots with the cameramen, ride the audio and microphone channels with the other hand, and that one person in the live director's booth, was me, M.D. Smith, IV. Later, we added a pair of studio lights in the director's booth and when the time came for "News Funnies of the Day" which came over the AP news wire (teletype machine), one camera would point through the glass panel into the director's booth, I'd stand up with a sport coat and tie, open up my microphone, fade the camera on me, then look up and deliver about 3 minutes of slices of life news stories. Then, fade back to the two-shot of talent in the studio, close my mike, open theirs, turn off those hot lights in a booth already too warm with all that tube equipment heating it up, and continue the show. We only had a 10-ton AC unit to cool the entire back part of the building where ALL of the equipment resided. Normally a lot was turned off, but during live productions AND in the summer time with every tube turned on, the heat was just about unbearable. I am talking high nineties even with fans blowing in every direction. When we had the money to do so, it was one of the first things we changed to a 20-ton unit. That was much better but still got on the warm side in summer with all those tubes.

I wrote commercial copy (as did others) and matched slides that may have been made by the salesman on location or copy for film shot on location, or slides made from the art cards by the artist. The station artist was a sign painter and did everything with brushes on different shades of gray 11" by 14" cardboard stock and used only a bottle of black and another of white poster paint. He mixed shades of gray for all other shades needed for logos, lettering, cartoon faces or whatever was not a B&W photograph. Everything being in black and white TV, colors didn't matter and could be highly misleading. There were two colors of red and blue that when seen in B&W were the same shade of gray and it was like a disappearing act.

When we eventually went to color, using colored paint, but we still had to preview the color cards live with our studio cameras on a B&W monitor to make sure it was good in B&W, since so many viewers still had B&W TV sets.

We only had a single, very expensive and difficult to keep running Ampex B&W VR-1000 VTR, short for Video Tape Recorder machine that used giant two-inch-wide reels of videotape for recording. It had hundreds of tubes inside the giant cabinet as well as two full height rack panels full of electronics and tubes.

When we first bought the station, commercials that ran with B&W slides that were made on location of a sponsor's store and merchandise ran in a slide projector into the film camera chain, where both 16mm movie film on two projectors were "multiplexed" with prisms along with a single slide projector with dual drums, so that while one slide was showing, the other would be changing and you would not see it on the air. Originally, the audio was read live by the "booth announcer" for every single slide commercial and station break from a "copy book" assembled in a 3-ring binder every day. Being in radio, we already used most everything on audio tape. So, the first thing, we had to record each commercial instead of having to read it live and put it on small reels of tape like radio, then the "switcher" would read along with the tape and push the slide change button at the marked spot where each slide change occurred.

Sometimes the switcher would lose his place, and get the slides out of order, or may have even loaded them wrong and the messed up the commercial to have the audio not match the pictures. This mess-up was called a "discrepancy" and had to both be noted on the commercial log that there was a problem, and then written-up on the discrepancy report what had happened, so the traffic and billing department would not bill the sponsor for that commercial, until a "make-good" had run that they had agreed to, thus fulfilling the contract number of spots for the month. If it could not be "made-good" because of a special football bowl game, well, we just lost the revenue, and the sponsor was not happy either. By 1965, we were using Spotmaster tape cartridge machines. The tapes looked very much like 8-track tapes, with an endless loop coil of tape inside the cartridge but were not the same. The beauty was that cart tapes were ready to play at the beginning of the commercial as soon as pressing the start button, would continue to run after the audio ended and automatically stop at the beginning of the spot each time, by way of sensing a "stop" tone put on the tape at recording time.

Later it was discovered we could add a second tone to the tone track of the tape, that could function for anything, and in our case, to change to the next slide. Now we didn't need the "copybook" anymore because when the cartridge tape commercial audio ran, it would change the slides in the dual drum slide projector exactly when it needed to change every time. Well, almost every time. There were still a very few mistakes, but nothing like before and manpower was cut way down.

So, any commercials that we could just run with slides and a companion tape recorder for sound, is the way a lot of commercials ran. The "Switcher" was in charge during station brakes between shows, to run these commercials. Some were 16mm film with sound, some were slide and tape, and a single Video Tape commercial might be mixed in the break as well. He sure was busy during that seventy seconds to two minutes depending on the length of the break. Then the switcher put everything back from where it came and started getting ready for the next break. During live shows or the news, that was every eight minutes or so. It was a quite demanding job and anything that messed up, it showed on the air and the switcher (now called "Master Control Engineer") was to blame. He was blamed even when the film broke on its own because we lost the commercial, didn't get paid and looked bad on the air, but it still happened occasionally. He always had a "trouble slide" ready in the slide drum which he'd quickly spin around and put on the air instead of a black screen and no sound, and at least the audience could see, "PLEASE STAND BY, WE ARE HAVING TECHNICAL DIFFICULTIES," which was the catch-all phrase for a lot of things that could and did go wrong.

The News Department:

It was a lot of work for a group of only four men and a part-time weatherman in those beginning years. The two newsmen worked together to decide what stories they planned to cover from that morning until about 3 p.m. when they had to return to process film. There was only one sound film camera simply called the "Auricon" which was the brand name and required heavy batteries of both low and high voltages to operate the tube audio recording system that put the soundtrack on the film of the camera as it was rolling. The other newsman and cameraman both had silent cameras and often went on separate stories to just get the film, to which we would add audio later or be read live by the newscaster that evening on the news. That was a bit tricky to add sound that way.

The black and white film was processed in a very slow machine that ran it through chemical baths at only a rate of four-feet-per-minute. Some days production would actually rip the film off the machine as it came out of the dryer door and put it on a projector. Later we got a B&W processor that was much faster at twenty feet per minute if I remember correctly. In 1968 we switched to color film & had to build a new room for the huge color film processor and bit vats overhead on two walls to gravity feed the many color, wash and fixer tanks.

At the end of the day, everyone went home tired. There was always more work to do, and things like cleaning up your workspace, area or desk were last on the list. As I look back on those years, everyone still enjoyed what they did and were part of a close-knit—and sometimes odd mix—a family that HAD to work together to pull it all off on the air on TV in those years.

The Network:

In 1964 the station was affiliated with the ABC Television network. There was this thing called "Network Revenue." Let me explain that and how it worked in those days.

When an affiliated station ran programs in Prime Time, which was 7 to 10 p.m., the network got paid a portion of the total cost of each TV market in proportion to the number of "homes watching television" the various markets averaged according to the Nielson ratings. In Prime Time, we got 30% of the network's charge for the homes our Huntsville market delivered to them. The "Network Hourly Rate" was the money that we negotiated with the network. Our rate at that time was $300 an hour, which meant we got about $90 IF all the commercials in a prime-time hour show were sold and ABC kept the rest. Other times of the day and weekend were less percent and sports had no compensation at all, but we had more local time slots to sell our commercials. That may not sound like an awful lot of money, but at the end of the month, it significantly added up for a young station trying to make a profit. It was about one-third of our total gross income each month. Network television 54 years later would be totally upside down with stations paying the network merely to have programming at all, and also paying the network part of their "cable fee" to the network. That's the fee local stations collect from the cable companies carrying the local stations and making a profit off of those TV signals and programs.

The Engineering Department:

Our engineering department could fix almost anything and often had to, since we could not afford the money to buy new equipment and sometimes, not even new parts. They even had to get some parts for the monster ole projectors machined in a shop since no replacements were available. We are talking some really old equipment we had inherited when we bought the station. One of our engineers was a young man my same age, and that would make him twenty-three years old for most of 1964, his birthday coming in November before he'd turn 24, and a month before mine in December. He was a young and very smart engineer who did not know the word "can't." There was always a way to accomplish a goal, and usually several ways and always a "cheaper way" to do the job. His name was "Cactus," a nickname he got at a very early age. His given name was Robert or Bob, but no one ever called him that.

Cactus even knew more about radio engineering than anyone we had on our payroll at WAAY-Radio, and soon he was Chief Engineer of both the Radio and TV stations. He would maintain that position for the duration of the Smith family ownership which ended in 1999 when I sold the TV station, having already sold the AM radio station many years earlier.

Cactus WAS the engineering department. He hired some very, very smart other engineers. Some had "quirks" about them, but they worked on the low pay side, and Cactus would "manage" them and isolate some of them from the rest of the staff because they needed "special treatment" of their talents and odd personalities. Cactus was, and still is as I write these memories, one of the brightest, smartest and most clever guys I have ever met in my entire life. Yet, he was and is so laid back and easy going with a country accent that you'd never suspect what is inside that skin. A lot of people over the years thought they could "get the best" of "Ole Cactus," but they simply didn't know with whom they were "messing." None of them ever did, business or personal.

I loved one of his expressions when someone was making trouble for the station, and he'd refer to giving them something more important to worry about, diverting the attention from us. He'd say, "If the flies are all over you, just put a pile of dog (stuff) in the other corner of the room to attract them." It worked most of the time.

Sometimes when one of his engineers did a really good job, or simply was underpaid but we could not come up with a raise, he'd give the person a gift, ranging from a free dinner for two up to, in one case, a new Collins ham radio Transceiver that he knew the man wanted and could not afford. But at the same time, even a raise would be small and used up each week, but this gift took the place of a raise for a year, and the engineer was delighted. He still talks about that bonus deal to this very day. Cactus was creative.

Cactus never fired anyone either. One of his expressions I loved then was that he "let them un-hire themselves." He would have some serious one-on-one talks, basically warnings. And if that didn't work, he'd "have the talk" and review previous conversations and in a lot of other words, tell them to look for another job right away. He would tell them their talents were special, but just not what we needed at our station, and while still having a job, they'd find other employment much easier. They always got the message, and soon they were gone. They didn't get fired, and they didn't exactly quit, they "just un-hired themselves" and found other employment. Of course, Cactus would keep me informed also.

Cactus is the man who introduced me to many of his favorite sayings he picked up over the years, much of it on ham radio talking to old timers. The two I liked the best are these:

If someone were very happy about something, he'd say, "There were smiling like a mule eating briars." That's about as wide a smile as I could picture. Another of his when a job was nearly completed, he would say, "Well, we're in the short rows now." I asked him about that, and he explained that a farmer's field was never square and when you were close to finished plowing the field, the rows he was plowing got shorter and shorter, thus, being in the "short rows."

Cactus also used one of our most favorite southern expressions when talking about someone who just could not cut the mustard at his job, "Bless his heart, he just couldn't hack it." Or Cactus might say, "Well, bless her heart, she just was dumber than a brick." It's a "Southern Sympathy" expression. Over the many years I have known him, he has put up with more kinds of crap from more kinds of people and yet finds ways to keep harmony amongst the employees, or his relatives and friends, "Bless his heart."

Cactus has real talent besides engineering brains. So, that's my tale of the engineering department in 1964 as well as Cactus over all the years to come that the Smith family would own the station.

Special Programs Department:

The weather maps of the day. They were sometimes drawn by the weather man who was just one of the announcers or salesmen at the time, trained to read weather basics. The weather maps were two large maps painted on canvas and covered with large sheets of Plexiglas and drawn on with "Magic Marker" refillable pens of different colors.

Black was used the most. After the maps were finished for the evening, the studio camera man, or whomever was available, took rags soaked with Lacquer Thinner which was the best thing to dissolve that permeant ink, and cleaned the board. To fill those big aluminum tube pens, you had to unscrew the felt tip about one-half inch in diameter and pour the liquid black in into it. That could be a very messy job and if it got on your clothes, it stayed there. I did the weather shows for a year in 1971 until we hired another man, and I can attest to what a mess it was. The current temperatures, humidity, chance of rain, etc., were menu board with white letters you pressed into the black felt slots.

Saturday afternoon at 5:00 to 6:00 p.m. was "The Ole Fisherman Show," with Joe Adams and Buddy Bingham (Johnny Evans) of Adams Distributing Company featuring Schlitz Beer. The opening and closing theme song, "Gone Fishin'" by Bing Crosby and Louis Armstrong, was perfect for the show.

The show comprised of a number of one-hundred-foot reels of B&W film we would process for them, and they would edit (or not and just run like it was) and narrate over the film telling of great fishing spots, lures, methods and showing the catch and locals that participated with them throughout the week. Sometimes it involved trips to Florida that were featured. Between films on fishing, we would run Schlitz Beer commercials which his distributing company sold and that program helped his business and brought a lot of revenue to our TV station back in the lean days.

Afternoon Kids Programming: "The Johnny Evans and Tex" show was a big hit. We had kids live on the show every day and a maximum of about twenty-five. One of them was usually celebrating his or her birthday. Johnny Evans had a ventriloquist doll dressed in a western outfit, he called "TEX." Johnny, likewise, was dressed in a cowboy outfit. Sometimes staff members would just call Johnny by the nickname, "Cowboy." He was also one of the very best salesmen over all the years he was there, because he was so personable and well known.

Morning entertainment 6-7 a.m.: "The Three Stooges." ABC programming started at 7:00 a.m. weekdays, and no one had even considered news in the sixties. I had small kids, and we owned unlimited rights to play "The Three Stooges" not only in our afternoon kids shows with other cartoons, but weekends and whenever. I did a slightly controversial thing of playing an hour's worth of them from 6:00 to 7:00 a.m. every weekday morning. It kept my small kids occupied while I was getting up and having breakfast and it also drew good ratings from not only kids but a fair number of adults as well.

The First Live Santa Show was in 1963. I still have to tell this true story of the First Santa TV Show ever on live television in Huntsville. It was on Channel 31 our very first year of ownership. My father had done them in the early days of Channel 6, WBRC-TV in Birmingham, and he wanted us to do one every afternoon and get a big sponsor who sold toys to buy it. I call this recollection clearly and it can be found earlier in this book.

A very small group of people did an amazing array of different jobs. You never got bored. You learned how to do many new things and we were always training backups for other positions. And, of course, people would leave of their own accord and we had to hire new replacements, and our small staff had to fill in for their jobs until a replacement was found and trained.

Part 3

WAAY TV 31 Challenger Video Tribute 1986

There was a tribute video that aired on Channel 31 WAAY-TV, Huntsville, AL on Jan. 31, 1986. John Jenkins of WMJJ Radio in Birmingham edited some of President Ronald Reagan's eulogy into a song of John Denver called "On The Wings Of A Dream." The powerful music was then illustrated by Cliff Windham of Ch. 31 with video from ABC, NASA and WAAY-TV files to make this powerful TV tribute. It aired January 31, 1986 (courtesy of M.D. Smith, IV).

The Challenger shuttle crew, of seven astronauts—including the specialties of pilot, aerospace engineers, and scientists—died tragically in the explosion of their spacecraft during the launch of STS-51-L from the Kennedy Space Center about 11:40 a.m., EST, on January 28, 1986. The explosion occurred 73 seconds into the flight as a result of a leak in one of two Solid Rocket Boosters that ignited the main liquid fuel tank. The crewmembers of the Challenger represented a cross-section of the American population in terms of race, gender, geography, background, and religion. The explosion became one of the most significant events of the 1980s, as billions around the world saw the accident on television and empathized with any one of the several crewmembers killed.

Ronald Reagan's Speech on the Space Shuttle Challenger Explosion (1986)

Nineteen years ago, almost to the day, we lost three astronauts in a terrible accident on the ground. But we've never lost an astronaut in flight; we've never had a tragedy like this. And perhaps we've forgotten the courage it took for the crew of the shuttle; but they, the Challenger Seven, were aware of the dangers, but overcame them and did their jobs brilliantly. We mourn seven heroes: Michael Smith, Dick Scobee, Judith Resnik, Ronald McNair, Ellison Onizuka, Gregory Jarvis, and Christa McAuliffe. We mourn their loss as a nation together.

For the families of the seven, we cannot bear, as you do, the full impact of this tragedy. But we feel the loss, and we're thinking about you so very much. Your loved ones were daring and brave, and they had that special grace, that special spirit that says, "Give me a challenge and I'll meet it with joy." They had a hunger to explore the universe and discover its truths. They wished to serve, and they did. They served all of us.

We've grown used to wonders in this century. It's hard to dazzle us. But for twenty-five years the United States space program has been doing just that. We've grown used to the idea of space, and perhaps we forget that we've only just begun. We're still pioneers. They, the members of the Challenger crew, were pioneers.

And I want to say something to the schoolchildren of America who were watching the live coverage of the shuttle's takeoff. I know it is hard to understand, but sometimes painful things like this happen. It's all part of the process of exploration and discovery. It's all part of taking a chance and expanding man's horizons. The future doesn't belong to the fainthearted; it belongs to the brave. The Challenger crew was pulling us into the future, and we'll continue to follow them...

There's a coincidence today. On this day 390 years ago, the great explorer Sir Francis Drake died aboard ship off the coast of Panama. In his lifetime the great frontiers were the oceans, and a historian later said, "He lived by the sea, died on it, and was buried in it." Well, today we can say of the Challenger crew: Their dedication was, like Drake's, complete.

The crew of the space shuttle Challenger honored us by the manner in which they lived their lives. We will never forget them, nor the last time we saw them, this morning, as they prepared for the journey and waved goodbye and "slipped the surly bonds of earth to touch the face of God."

Dan Akens - WAAY & WNUE Radio Days

WAAY
1550-HUNTSVILLE
Radio

WAYS OFFICIAL
FAMILY SURVEY NUMBER 31
WEEK OF AUGUST 3, 1959

1. A BIG HUNK OF LOVE/MY WISH COME TRUE
ELVIS PRESLEY — RCA
2. TIGER
FABIAN — CHANCELLOR
3. CRACKIN' UP
BO DIDLEY — CHECKER
4. JUST A LITTLE TOO MUCH/SWEETER THAN YOU
RICKEY NELSON — IMPERIAL
5. WHAT IS LOVE
THE PLAYMATES — ROULETTE
6. SINCE YOU'VE BEEN GONE
CLYDE McPHATTER — ATLANTIC
7. LONELY BOY
PAUL ANKA — ABC PARAMOUNT
8. SEA OF LOVE
PHIL PHILLIPS — MERCURY
9. RAGTIME COWBOY JOE
DAVID SEVILLE & CHIPMONKS — LIBERTY
10. I'LL BE SATISFIED
JACKIE WILSON — BRUNSWICK
11. LAVENDER BLUE
SAMMY TURNER — BIG TOP
12. WHAT A DIFFERENCE A DAY MAKES
DINAH WASHINGTON — MERCURY
13. DANNY BOY
SIL AUSTIN — MERCURY
14. HIGH HOPES/ALL MY TOMORROWS
FRANK SINATRA — CAPITOL
15. PERSONALITY
LLOYD PRICE — ABC PARAMOUNT
16. THERE GOES MY BABY
THE DRIFTERS — ATLANTIC
17. SWEET SUGAR LIPS
THE KALIN TWINS — DECCA
18. I'M COMING HOME
MARV JOHNSON — UNITED ARTIST
19. SINNER MAN
TOMMY SANDS — CAPITOL
20. THE WONDER OF YOU
RAY PETERSON — RCA
21. PURE LOVE
SONNY JAMES — CAPITAL
22. SMALL WORLD
JOHNNY MATHIS — COLUMBIA
23. CROSSFIRE
JOHNNY & THE HURRICANES — WARWICK
24. A LONG WAY FROM ST. LOUIS
PEGGY LEE — CAPITOL
25. PARTNERS
JIM REEVES — RCA
26. STAY
DEE ARLEN — COLUMBIA
27. YOU ARE IN LOVE
PERRY COMO — RCA
28. MY HEART IS AN OPEN BOOK
CARL DOBKINS JR. — DECCA
29. THIS I SWEAR
THE SKYLINERS — CALICO
30. SEE YOU IN SEPTEMBER
THE TEMPOS
31. SOFT SUMMER BREEZE
EDDY HEYWOOD — MERCURY
32. THE WAY I WALK
JACK SCOTT — CARLTON
33. I'M GONNA BE A WHEEL SOMEDAY/I WANT TO WALK YOU HOME
FATS DOMINO — IMPERIAL
34. HUSH-A-BYE
THE MYSTICS — LAURIE
35. CHOSEN FEW
FOUR LADS — COLUMBIA
36. THANK YOU PRETTY BABY/WITH ALL MY HEART
BROOK BENTON — MERCURY
37. FIVE FEET HIGH AND RISIN'/I GOT STRIPES
JOHNNY CASH — ATLANTIC
38. IF YOU LOVE ME/SO HIGH SO LOW
LAVERN BAKER — ATLANTIC
39. I'VE BEEN THERE
TOMMY EDWARDS — MGM
40. CRY
KNIGHTSBRIDGE STRINGS — TOP RANK
41. BLUES STAY AWAY FROM ME
BILLY VAUGHN — DOT
42. HEY LITTLE GIRL/A GIRL LIKE YOU
GARY STITES — CARLTON
43. ONLY WISHFUL THINKING
PETER KIM — CLICK
44. TIME MARCHES ON
ROY HAMILTON — EPIC
45. BIG SURPRISE
FOUR PREPS — CAPITOL
46. SOLDIERS JOY
JERRY REED — NRC
47. TWICE AS NICE
CLYDE MCPHATTER — MGM
48. SLOW MOTION
WADE FLEMMONS — VEE JAY
49. LOTS OF LUCK
THE HILLTOPPERS — DOT
50. THIS LITTLE LOVE OF MINE
ED TOWNSEND — CAPITOL

---

51. WHAT'D I SAY (Part 1&2)
RAY CHARLES — ATLANTIC
52. LINDA LOU/RED SAILS IN THE SUNSET
RAY SHARPE — JAMIE
53. JOHNNY ROB
JOHNNY HORTON — COLUMBIA
54. THE MUMMY
BOB MCFADDEN AND DOR
55. CHECK-MATED AND BINGOED
BILLY AND DON HEART — ROULETTE
56. THE THREE BELLS
THE BROWNS — RCA
57. I'VE COME OF AGE
BILLY STORM — COLUMBIA
58. WITH MY EYES WIDE OPEN
PATTI PAGE — MERCURY
59. WE BELONG (TOGETHER)
JIMMY FORD — STYLO
60. KISSIN TIME
BOBBY RYDELL — MGM

"WAY" STARLIGHTERS
SPOTLIGHT SONG HITS

DAN AKENS
THE SHAPE I'M IN/YAYA
JOHNNY RESTIVO — RCA

DICK ESSNER
WALKING ON AIR
JIMMY BOWEN — ROULETTE

JERRY LEE
I'M GONNA GET MARRIED
LLOYD PRICE — ABC PARAMOUNT

DEE SCOTT
TALL COOL ONE
THE WAILERS — GOLDEN CREST

1550 YOUR RADIO "WAY" IN HUNTSVILLE

Above is a copy of the Weekly "Top 40" Hits from WAAY Radio in Huntsville, Alabama, Circa 1959's. Also, a closer photo of the DJ group at the bottom of the list.

Jerry Lee (Real Name Jerome Tisdale) went on to the recently Purchased Smith Broadcasting station in Fort Walton Beach, Florida. WNUE (New Radio)

At about the same time Dan Akens went to the Smith Station in Tullahoma, Tn. WJIG RADIO. While there, Dan, who was in his mid-20's was located by his "Draft Board" in Illinois, and summoned to the Induction Center in Montgomery, Alabama. (It was in between the Korea and Viet Nam wars, so, the only fighting was trying NOT to go to Montgomery!)

Anyway, Dan lost and was sent to Fort Jackson for Basic Training and then to Fort Gordon for "Radio Relay, & Carrier Operator Training." (I picked that because it had "Radio" in the title.)

Ooops...it really meant I was being sent to a Mountaintop in Korea to "Turn On & Operate" radio relay equipment every morning!

So, while I was in Oakland, California waiting to be shipped out, the war ended, and I was "Mustered out of the Army Signal Corps," and sent by Greyhound bus across country to Fort Walton Beach, Florida, where MD Smith and his wife Kirby had purchased WNUE RADIO.

HERE's where it gets "semi-interesting"...tragic...and, changed my life! For those paying attention (or, still with me) the aforementioned Disc Jockey Jerry Lee, had gone down to WNUE Radio (which was closer to his Hometown of Crestview) AND another friend of ours.

Bill Dewey also went down from WAAY to WNUE.

This is where a tragic event changed my life! Jerry and Bill were swimming in the clear, green water just below the Destin Bridge at the Destin Pass. (Where all the fishing boats and pleasure cruises came in from the Gulf.)

Tragically, Bill Dewey got caught up in an undertow. Jerry tried to save him but could not! And Bill drowned. Just out from the jetty at the Destin Bridge!

Both MD & Kirby knew I was getting out of the Army. My old jobs in Huntsville and Tullahoma were filled. So, they offered me the job of Deejay, and what became Operations Manager for WNUE AND LATER for Smith Broadcasting in Birmingham.

But, wait...THERE'S MORE!

Kirby Spent a good deal of time at WNUE. (Since they had vacation home on the Gulf, and she also did a lot of work at the station.) So, one day Kirby said, "Since you don't know anyone here, I've got just the girl for you, a family friend's daughter, Diane DeRoche."

So, Kirby set me up on a blind date with Diane. However, after it was set up, I got a call at the Radio Station while I was "on air" as a DeeJay. Diane said she had a "family conflict" and couldn't go, BUT, A GOOD FRIEND OF HERS, Martha Totten, had agreed to go and had promised not to "move in on Diane's territory."...AND, THE REST IS HISTORY! Love at first sight! Beautiful, sweet, smart and, truly one of the nicest people I had EVER met!

Best Regards, "Dan, the Medicine Man—with, Music & News to Cure the Blues!"

Dan Akens—One of Smith Broadcasting's Pioneer Dee-Jays at WAAY RADIO in Huntsville, Alabama. Asst. Mgr. at WJIG In Tullahoma, Tn....and, DeeJay/Pgm. Dir. at WNUE.

In Fort Walton Beach, Florida. Then, later as operations manager at Smith Broadcasting's Production Headquarters in Homewood, Birmingham, Alabama. Dan salutes the Hundreds of Smith Broadcasting Alumni and fondly remembers "MD 3", The Lovely Kirby and "MD IV"!

Bettie Higgins' Stories, Cactus, also '84 Winter Olympics

The photo here in this section is from the TV station's January 1983 "Winds Of War" press party. There many of our staff dressed in WWII military outfits to set the scene of this multipart ABC TV event on Television. It was a huge event and one of the largest rating Special Programs ever shown on 31 TV. It was a huge success, and Bettie Higgins was the driving force behind all the details and many of the ideas to make it special. Kudos to her for the part she played. [Editor, M.D. Smith, IV]

Perhaps others might enjoy this story…

During the days when so much commercial production was done at 31, we had clients who came to the station to view their spots. The nice conference room had been built, but the monitor in there left a lot to be desired. So, I told Cactus we really needed a spiffy new monitor for presentations of spots to clients.

A few days later he appeared in my office: "Miss Bettie," he said, "got you a great monitor in the conference room. It's an itchy-bitchy!" I just stared at him and said, "What?"

He said, "That's the brand—itchy-bitchy—it's a great one."

I said, "Cactus, are you trying to say Hitachi?" "Naw," he said, "Of course I can say Hitachi—this one's an itchy-bitchy."

He followed me out of my office and to the conference room where I saw a big, beautiful MITSUBISHI. But, of course, the monitor was forever and always "The Itchy-Bitchy!"

Here's the Bob Sullivan story . . .The Wizardry of Bob Sullivan

In 1984, when the Winter Olympics were held in then-beautiful Sarajevo, Czechoslovakia, ABC was really hyping it, and I was trying hard to help our Sales Department in selling various components such as the Luge Competition. Johnny Evans strolled into my office one day and said, "How can I sell the 'Luge Competition,' Bettie, when I don't even know what the hell a luge is!"

All in all, it was tough sledding (pun intended) selling the Olympics of 1984—far away; re-broadcast, often late at night…so, when I got a call from ABC one afternoon asking if we would like a one-on-one interview with our Sports Anchor, (Bob Labbe) and Jim McKay from Sarajevo, I said, "absolutely!" "Fine," Tom said. "I'll call you back with date and time for taping."

I immediately went to Engineering and sat down in front of Bob Sullivan. With great enthusiasm I explained to him about the interview with McKay and Labbe…he listened, the ivory cigarette holder, jutting upward from a corner of his mouth and the burning Pall Mall emitting clouds of smoke. Finally, when I stopped for breath, he removed the holder and said words I had never heard, and never expected to hear from Bob Sullivan. He said, "I'm not sure we can do that." Bob never said we couldn't do something…he always did whatever was necessary …he made miracles happen, and now I really needed one!

I didn't say anything and for a long minute he just puffed the Pall Mall and looked thoughtful. Finally, he said, "Well, let me work on it—I do have a new pair of alligator clips." And he smiled. My world started spinning again!

All was well until the next day when Tom at ABC called me back. He began on an ominous note, "Um, um, I think we have a problem." "No, we don't," I said, "we can handle this." He said, **"Well, my ruler was on the wrong line on my computer print-out, and I thought you**

were in Indianapolis. We can't offer the feed to a market as small as Huntsville."

I don't remember exactly what I said, but in my best **"peel the wallpaper off the wall voice,"** I pointed out that a deal was a deal and there was no way ABC was going to renege on this deal—no way, Buddy!!! I think I recall asking for Roone Arledge's phone number!

All went well; Bob worked a miracle, Labbe talked to Jim McKay, and I found a picture of a Luge sled for Johnny Evans!

Bettie Higgins

Dean Abla – Sales & Motorcycle Wreck Meeting the Manager

An interesting thing happened to me on my journey to WAAY-TV. It was in December 1970 and I had just been hired by Dan Whitsett to join the Sales Staff. Since I was to report to work effective January 1, 1971, I took my family for a ride up the mountain on Christmas Day to show them where I was about to begin a new career.

Just as we rounded the hairpin curve from Bankhead Parkway to Fern Drive, here came down the mountain some guy, who had obviously had too much Christmas cheer, on a motorcycle. He lost it and wound up in the ditch in that curve! I stopped my car and got out and rushed over to see if I could help him. It seemed the worst catastrophe was his smashed flask in his hip pocket. We were successful in pulling his bike up out of the ditch and he managed to climb back into the saddle and head BACK UP the mountain.

On Monday, following Friday's New Year's Day, I reported to work and was extremely excited about my new job. Everyone was very pleasant and friendly and I was enjoying all the introductions. Finally, there was a man with his arm in a sling who was introduced to me as

MD Smith, IV, the general manager of the station. It only took a couple of seconds to realize that he was the guy on the motorcycle!! His shoulder had been dislocated in the accident! As we laughed and talked about the incident, he told me the worst part was that, on his way back home, every time he would stop, he forgot to put his foot out and his motorcycle kept falling over!!!!! I think MD sold his bike soon afterwards. *[Editor's note: Yes, and he has not owned one since then.]*

The 31 1/2 years I worked at WAAY-TV were extremely rewarding. I met lots of really fine folks who passed through those doors, some of whom still work there. I also was privileged to work with and for some great clients who became friends as well. I appreciate the company's willingness to allow me to have membership in the Advertising Federation and the opportunity to give back to the community which was so good to me all those years. I must admit that the first 29 years were the best and that working for the Smiths was a rewarding span in my lifetime.

Terry Heaton – The Smoking Room

While I have hundreds of fond memories of the old 31, nothing warms my heart like thoughts of the smoking room, that place where political correctness ended and liberty began. I don't smoke anymore, but I did when I worked at WAAY.

The smoking denizens gathered like clockwork at certain times of the day. There must've been some magic signal that was transmitted via the cosmos to the minds of smokers throughout the building, because we all seemed to arrive together. In fact, the worst thing was to arrive as others were leaving, usually laughing, because you'd obviously missing something funny. We laughed a lot back there. We cried some too. We were a little community within the overall WAAY community. Smokers are like that.

People put up with my cigars, but smokers are like that too, tolerant to the max. I'd light one up in the morning and leave it in one of the WAAY designer ashtrays to be relit throughout the day. Only once did it disappear with the trash before I was finished with it. My friends saw to that. Smokers are like that too.

It was our lunch room and the heart of the WAAY gossip wire. Nobody was off limits, not even every past or present member of the Smith family. It was a time and a place to embellish, to boast, and mostly to let off a little steam. And who can forget the smoker (no names, please) who fell asleep and set the couch on fire. Or the noisy fan. Or the times the smoke was so thick, it was hard to breathe (did we really care?) Or the complaints from non-smokers when one of us left and carried the fragrance with him throughout the building.

I spent times alone in that room too, thinking about the day's news or some future project. When the weather was nice, I went outside to enjoy the view, but there was something special about smoking in that room. Perhaps it was the knowledge that we were living on borrowed time, that the curse of political correctness would one day come upon us and we'd be forced outside like most of the other smokers in the country.

How could I not enjoy working for a guy who actually let me smoke in a comfy place throughout the day? This, I believe, was another of the many acts of kindness and grace granted to the family of Smith Broadcasting employees.

Cliff Windham – News & Tornado of 1989

I was certainly much younger and slimmer when this picture was made in 1983 or 84! Just a year or two earlier, in 1982, I had come to WAAY-TV, from WTVM-TV in Columbus, GA where I spent 5-years producing, reporting, and assignment editing. That's where Mark McGee and I first worked together, before I encouraged him to join the news team at WAAY-TV. Before that, my TV career started in my hometown of Atlanta as an intern at WXIA-TV, while a student at Georgia State University. At WAAY-TV I enjoyed 14 years of working with M.D. along with many others whom I rank among the most talented and professional people I've ever known.

I'm now proud to be associated with an equally fine staff of people at Baron Services in Huntsville, where I'm a Regional Manager involved with marketing our newest desktop weather monitoring and storm tracking systems called Threat Net™ and Mobile Threat Net™ that are used by Public Safety agencies and other organizations and businesses, across America. Company President and CEO Bob Baron was the Chief Meteorologist at WAAY-TV while I was News Director there.

He jokes that I used to be his boss, and now he's my boss...with a vengeance! As you probably know, Baron Services is also the leading manufacturer of Doppler radar Systems for TV stations in the U.S., like WAAY-TV.

What I'm doing in this picture is helping to decorate the 31 float for the WAAY-TV Christmas Parade. This annual ritual would take place over a few cold December nights in an unheated warehouse downtown. Hot pizza and coffee offset the chill. (Un-huh, notice the bottles of beer in the background?) There were usually 20 or 30 people from the station, including some family members, who like Santa's elves busily stuffed thousands of multi-colored tissue paper poms into a chicken wire frame stretched over a flat-bed trailer. Before we knew it that rusty trailer covered in chicken wire, was magically transformed into a holiday delight. At the same time many other people across the Tennessee Valley were doing the same thing, as they assembled their floats for the parade.

Many times the WAAY-TV float would be pulled by Jamie Cooper, who was the "Country Rover" feature reporter at the station. Riding on the float would be news anchors and other staff members and their kids. I recall the long hours folks like Bettie Higgins and Debi Bradford spent coordinating the thousands of details involved in making the parade, with its scores of entries, the big success it always was. I remember they had to be especially sensitive to the placement of the many entries featuring horseback riders. Since the horses tended to drop little Christmas presents of their own along the route, the goal was to not bunch these entries together, or the non-horseback entries following behind would have literally become bogged down in a sea of "presents."

There were also many serious times to remember. One of the most significant and devastating events in Huntsville's history was the Tornado of November 15, 1989. I've attached a Huntsville Times article from November 17, 1989 that described how the local media responded to this event. It also points out that, thanks to the foresight of M.D. Smith, WAAY-TV was the only Huntsville station with a generator that allowed it to continue broadcasting while the other stations were dark. I recall how virtually everyone one of us at WAAY-TV worked virtually three days without sleep as we covered the rapidly unfolding events surrounding this story. We produced a two-hour documentary, called "Tragedy, Turmoil and Triumph" that show highlights of that dramatic news coverage. **We sold thousands of VHS copies of this program with proceeds going to the American Red Cross Disaster relief effort. It raised $26,000.**

A few months later, a check was presented by M.D. Smith, IV to Richard Matthews of the Red Cross. I still have one of these original VHS copies. M.D. sent a note to the staff thanking everyone personally for their efforts and provided, as I recall, **50 silver dollars** to everyone as a bonus.

My best wishes go out to each of you, and your families. I hope to see many of you at the reunion. I'd also like to say thanks to M.D. for pulling this all together.

Ed Hedden, Ph.D. – Art Director & "Batman" on a Harley

[Editor's note: This photo of Ed was when we got one of the original Batman costumes from ABC and tried to make good use of it. We shot some ID's and even a Burgeron's Kentucky Fried Chicken commercial with it. But we needed someone who fit the costume and would agree to wear the tights. Ed, our artist, agreed. So we have a number of photos of "Batman" Ed Hedden in the files.]

[Ed has been in contact and says he will be in Huntsville and attend the 2003 reunion. I look forward to Ed sending a few stories to go with the photo. Ed has now earned his Doctorate and thus entitled to add the "Ph.D." to his name. I am sure he is proud of this accomplishment.]

Hi all,

As Art Director at WAAY-TV from November of 1965 to September of 1969 I had the pleasure to experience the wild transitional days of television. The least of which was going from black and white to color. There were many events that still seem like they happened only last year. I vividly recall Tiger Terrell Whitaker go-cart entrance taking out one tripod leg of the brand new Plumbicon camera that I was operating. The deafening silence, the gasps, and Cactus Gay running out to check the condition of his new toy. The thing was very heavy especially from the back side and leveraged position. I remember doing everything that I could to keep the camera from smashing onto the concrete floor. Fortunately, a bent lens frame was the only damage. By-the-way, no one was interested in my condition at the time.

In later years, at the PBS station in Illinois that I retired from, we built a Big Bird costume. The experience at WAAY-TV wearing the Bat Man outfit cured me of ever getting into another costume. I can still smell the BO and will never forget how hot and uncomfortable that experience was. The broadcasting facilities must have tripled in size during those years. There was constant construction, new equipment, and moving. Who can ever forget Sonny Limbaugh telling his wild tales about doing everything. The most repeated story was how he would come in early in the morning, start the transmitter, set up the news set, turn on a camera facing the spot where he was to sit, rip the latest news from the wire, go to the control room and bring up the camera on an empty set, than walk in, sit down, and read the news.

His exit stories were just as interesting. All this was live television during those years and everything that could go wrong did at some point. The talent was amazing at cover-up and recovery. Talent at PBS was just as interesting and difficult. Some of the strangest people in the world work in television. I recall one disgruntled sales person spending his last night inserting four letter words into all of the commercials that he had made. It was a real mess re-cutting them the next day. We produced a film representing the hundred-year history of men's clothing. In college it had been explained to me that at sixteen frames per second no one could distinguish a single frame. I had also read a book called "Subliminal Seduction" so two Play Boy center fold pictures were included. One near the beginning and a different one near the end. When the advertisement ran the second day some irate preacher called MD and the film was reviewed.

The order was to splice it out immediately. It was tempting to leave the second picture in but we cut it as well. I don't think anyone ever knew about the second frame. It's still a mystery to me how anyone could have seen a single frame. I still have a few copies of the many slides that were made. If I can locate them, I'll donate them to the cause.

The artist (sign painter) that proceeded me (Howard Troutman) was from Paducah, KY only eighty miles south of Carbondale, IL which became home for thirty years. I think MD III thought that all artist were sign painters. The order came in to paint the News Hound on the tail of the company airplane (see the Cessna 172 in the history section), another experience that will never be forgotten. With a little touch-up from another sign painter it turned out OK. The whole Smith family were avid art critics and seldom did a week go by without some change to the on-air artwork produced. MD IV was a great supporter and bailed me out of tight spots many times.

The availability's chart that we produced was a big hit and was adopted across the country. It's too bad that we did not copyright it. Eating lunch at the country store next to channel 19 was another fond memory. The Lamar low carb diet event was an important lesson learned. Having rocks rip through the top of my MG from blasting at channel 19 was interesting. They were always trying to catch up to channel 31. Over the years I've missed the many good friends and exciting times that we had at WAAY-TV. I look forward to our reunion.

Ed Hedden Troy, AL

Mike Sullivan – News 80s, Bad Live Newscast & strange food

Mike Sullivan, Bettie Higgins and Mike's wife, Patty in photo at "Winds of War" Press Party 1983

I was news director at WAAY-TV from Sept. 1980 to March 1984.

Memories:

Words of wisdom from Dan Whitsett

I remember one time during my tenure at Channel 31, our news ratings dropped slightly. You had made the decision that even though the numbers had dropped, the sales department would continue to maintain the same rates. We were discussing it in the weekly department head meeting. You explained your intention and asked Dan to tell us what he planned to do about it. Dan said he and Redge Swing were very concerned because they didn't know how they could sell at that same price. But then he continued, "Then Redge and I looked at each other and realized, heck, if it was easy anyone could do it, and they wouldn't need us!" That thought has stuck with me through some tough goings over the years.

Midnight Audition

When I arrived in the fall of 1980, we were, as they say at my alma mater, "hurtin' gators." An exodus of key news personnel left the news dept. with just the bare skeleton of a staff. Brenda Wood had left. Dave Stanley was on his way out. LuAnn Cahn left the Friday before I arrived. Several others, whose names I don't recall had also just recently left and had not been replaced. One of the first goals was to hire a weeknight co-anchor team. (Bob Baron and Rick Davis were still in place in weather and sports.) We hired Jim Marsh from Lafayette, La, but were having difficulty attracting and hiring a female co-anchor. Bruce Northcott, from Magid, came for a visit and asked if we had looked at the weekend "weather girl" at Ch. 48. (At the time, even considering hiring someone from Ch 48, let alone a weekend weathercaster for a front-line anchor job, was heresy.) We had no idea how this person could read news, so we contacted her and arranged for a "midnight audition" at the station. We literally waited until almost everyone had gone home one evening and snuck her in with a minimum crew, sworn to secrecy. As it turned out Kelly Cooper did a fantastic job and we hired her. She was a very instrumental part of our climb back to #1 before she left us for a station in Austin, TX about two years later.

Parades

The WAAY TV-Radio All Volunteer Christmas Parades were always big events when I was there. We were always torn between putting our front-line on-air talent into the live broadcast or to put them on the station float. One year (1982 or 83) we decided to put all the key anchors on the float, so it was left up to Liz Hurley and myself to anchor the 2-plus hour-long parade broadcast. We knew it would be interesting, seeing as neither of us had ever done a parade before. And I wasn't even a regular on-air person. As it turned out, we had a terrific time doing the show and no one threw any rotten vegetables at us for the effort. I think Carl Spurlock directed. It was a wonderful experience.

The live show that wasn't

In the summer of 1981, the Huntsville TV market was turned upside down when Ch 48 conducted a very successful talent raid on the other two stations, snaring the front-line news anchor team of Tom Kennemer and Missy Ming from WHNT and our sports director, Rick Davis. In the end, we probably benefited from this event more than anyone else. While we missed Rick, the loss of Tom and Missy chopped WHNT off at the knees. Unfortunately for Ch 48, the addition of these three doubled their ratings, which were pretty small to begin with, but still did not make them a serious competitor in the ratings competition.

However, at the time it happened, we were all very concerned. It was decided at the last minute, we would do an entire week's worth of road/theme shows the week the new team premiered at Ch 48. We started off with a couple of easy ones, like from the Space and Rocket Center and from downtown. Then, around mid-week, we planned to produce our entire newscast from a boat in the Tennessee River. We had some concerns about sending our entire on-air team out of the station with no back up. MD and I had a lengthy discussion about this. He was quite adamant that we should put all our eggs into this basket. I think the term "no guts, no glory" probably came up a couple of times. He said, "Mike, if there is something you just absolutely have to do and you have no other option, then you'll do whatever it is you have to do to make it happen. So just go do it."

We tried to cover every possible scenario. I think we even had a spare generator on hand. I'm sure you probably know what happened next. Fast forward to the evening of the telecast. At about three minutes to air, the microwave signal from the river failed. As it later turned out, a "diode" (I still don't know what a diode is) failed in the microwave transmitter. Even with Cactus, Bob Sullivan and most of the engineering staff on hand, there was nothing they could do to get it repaired in time. In the newsroom, I grabbed my coat and tightened my tie and ran to the studio. The worst wasn't over, because we had no weather, no sports and only about five minutes of content, the rest having been planned to be live interviews and such from the river. I started the show, praying that the microwave would come back to life. In the meantime, Helen Howard (Carroll) started bringing me scripts, wire copy, network packages, anything to fill time. We "packed and filled" for 30 minutes until, mercifully, we finally managed to get to the end break. The viewers probably saw the worst excuse for a newscast in the history of Huntsville television, but from studio on top of the hill, we were just glad we got through it. *[Editor's note: "And hey, Mike, we got all the spots run in the show and the sponsors paid for them."]*

Elections Ahead of Their Time

Before the days of computer/character generator interfaces, election coverage was an interesting proposition. It was virtually impossible to manually enter numerical results into a Vidifont or Chyron fast enough to keep up with the activity. In 1982 (I think) we joined a consortium of other Alabama news media to run a statewide tabulation system for key races. It would be delivered to the station via a computer. The problem still remained, "How do we get the numbers on the air." Someone, I believe it was Bob Sullivan, came up with the idea of using a computer monitor with a "video out" port. We ran a coax from the computer monitor into an input at the switcher. At that point we only had white characters and numbers on a black screen. However, by using color backgrounds and keying the numbers that came from the computer, we ended up with a pretty-darn-good looking election graphic. As a result, we were able to put the latest results directly from the computer and onto the air without any delay. Today that kind of capability is just taken for granted. But in 1982 it was pretty radical and demonstrated a lot of ingenuity.

Interview Lunch

When I went to Huntsville for the job interview, we had lunch at your house, MD. You were nearing the end of an incredible weight-loss program and were about to run in your first marathon. When we went to your house, we were standing in your kitchen and you asked me what I wanted to eat. I said a sandwich would be great, and you gave me the ingredients to put together a ham or bologna sandwich. Then, while I watched, you took an entire can of tuna and emptied it on to a piece of whole wheat bread. You laid a few sardines across the top of the tuna, added some chopped onion and sprouts, and topped the entire thing off with picante sauce. (I may have confused a few ingredients. It was 23 years ago. But the general picture is accurate.) *[Editor's note: "What a great memory, Mike, that's exactly the things I was eating on the Weight Watcher's Diet."]*

I just remember as we were sitting down at your kitchen table thinking to myself, "I'm glad when he asked me what I wanted to eat, I didn't just say, 'whatever you're having, MD.'"

Tim Clemons – Director, "Mock News", Quad Machines, and more

Tim Clemons, Producer-Director 1982-1998

I can't tell you how much I've been enjoying all the stories about WAAY-TV. Even those from the people I've never met. 16 years there was quite a haul, so I have quite a few stories, some I'll mention below.

I started out at WAAY in September 1982. Interviewed with Lew Koch and took the proverbial "spelling test." I recall calling Lew "Mr. Koch," pronouncing it like the former New York City Mayor's name. Lew said, "Ah, well, call me Lew."

My first job: Studio Supervisor. Sounds pretty impressive doesn't it? Job description: Sweeping the studio floor, ordering/changing lights in the fixtures (trained by John Keel), running camera in the studio (putting art cards on easels for John Stanners, Carl Spurlock and Jeff Davis), and the highlight each day, running newsroom camera for the 4:00pm news brief. Schedule: 1:00-4:00 each day. The Big Money! But a foot in the door!!

I was later assigned to other part-time positions…running camera for the news (weekends and weekdays), VCR operator (which included loading over-the-shoulder photo slides in the projector), Vidifont (aka character generator) and audio.

First fulltime job, Master Control (switcher). Weekend sign-on. Trained by Steve Kerr, one of my fellow "Billie's Bandits" (translation: anyone who attended Calhoun Community College with instructor Billie White). Later ran MCR during weekdays and primetime. I remember whenever lightning would strike the tower, which was quite often, every light on the switcher buss would light up like a Christmas tree. You had to re-punch your original source on the air to get the picture back. I also remember the old G.E. transmitter, which we called the pizza oven due to its appearance. And, the transmitter on/off switch on the wall held in place by a jar lid…no, really…a real lid from a jar! Mark Smith took that with him the day the transmitters were sold and hauled away. I also remember loading, and babying, those huge quad machines. Remember the cue-track? Those of you who know what a cue-track is will remember Sharon "Jamaica" Sandford counting down in her strong Jamaican accent…5-4-3-2-1. And the "dub masters" built by Harry Hatfield for the news. There were no commercial dub masters during primetime. You had to run each commercial and station i.d. separate, sometimes using 3-4 different quad machines, and each with its own 5-second pre-roll! Now THAT'S adventure switching! Harry always built masters for Liz Webb. The guys, we were on our own. Also had to load programs on the old 16mm film projectors. These days, switchers have cart machines and servers. They don't know how lucky they are.

Began directing the 6 and 10 news in 1984 under the watchful eye of Mark Smith, whom I still refer to as my mentor. We even "staged" some newscasts in between the 6 & 10 so I could practice, later referred to as "Mock News". I've still got some of those clips with John Hill, Gary Dobbs ("this hurricane's so strong, these palm trees are losing their nuts!"), Bob Labbe and Heather Burns…reading the news in her best southern-belle drawl. Absolutely hilarious! Mark and Allen Dunkin were both about to move back to Selma/Montgomery. When they were gone, I was directing ALL newscasts, seven days a week. After about six weeks of this, Lew had me train Liz Webb at directing. Funny, less than two years earlier, Liz was training ME to run camera!

I also remember directing the news the night of the '89 tornado. Kip Cole and I shared directing duties from 4:30 that afternoon all the way to the next morning when the a.m. news crew came strolling in. What a sad night for the city, but a real badge of honor for the crew and management of 31. Every employee later got a bonus check for all the hard work from that night.

After eight years of directing, I'd about had it with punching the buttons of that old GVG-1600. Lew started letting me do some producing around 1991 (PSA's, etc.). Just some little things to cut my teeth on. Eventually got out of news altogether and was given the position previously held by Deb Nepolitano. Before all was said and done, the crew and I racked up over 30 Addys...Golds, Silvers, Citations and one Best of Show.

And let's not forget the WAAY-TV Christmas Parade. Yep, produced and directed that too, for 13 long years. Through rain, sleet, snow and hell. But it truly was a blast and excellent experience.

One very memorable practical joke: You may have already read about my famous "loop carts" where I would record myself in the audio booth saying nothing but "Hey!" in a very calm voice. This would loop and replay every 60 seconds or so. We would pipe this audio out in the studio and really freak some people out. Another loop cart event was the time I recorded sound effects of chickens. Before leaving for the night, I left it playing into the studio. Gary McClendon, a sign-on switcher who drove every day from Albertville, came waltzing through the studio and heard chickens. He then proceeded to call Cactus in Scottsboro at 5:00 in the morning saying "Cactus, there's chickens in the studio and I can't find them anywhere!"

Who could forget all the times we had to "check the transmitter" (wink, wink)!

And, the "31 Staff Cars." Some may recall these old cars we bought from Mr. Earl and Adrian which we proceeded to trash and park along the side of Monte Sano Boulevard (we would move the cars to a different spot each day) one with the word "downtown" with an arrow spray painted on the door. On the other side, "lost our lease, moving soon!" anD "Eat More Possum" tag on the front, and "31 News

Everywhere" stickers all over for good measure. They were eventually towed away by the City.

Left WAAY-TV in 1998. I just felt it in my bones that things were about to really change at 31. About a year after I left, the station sold to GoCom.

Thank you, M.D. and Lew, for all the opportunities and all my friends for making "the 31 years" the best times of my life. Oh, how I miss it.

Gary Dobbs – Weather "The Weather Dude" & Credibility Building

The Hiring…

I had been at WRSA, beautiful 97 radio, for three years when one of our part time employees, Alan Duncan, approached me about submitting a video resume for replacement of the soon departing Bob Baron, the undisputed king of weather in the market. Alan was also part time director/producer at 31. My background included a stint as sports/weather director at WCFT CBS-33 in Tuscaloosa back in 1969-72. My background also included years as a shoe salesman and a factory rep. for Honeywell Corp. But, given I was wanting to get back to television, I decided to put together a tape for MD viewing. Alan brought me in to the 31 studios late one night and taped me doing a weather segment (and the first time I'd ever seen a chroma-key wall). I received a call from MD a few days later and he wanted to talk. It seemed more like a courtesy thing in that first meeting and as I recall he assured me he was looking at some of the top applicants in the country. I assumed my chances were pretty much nil. But he wanted me to cut another demo tape; so he took me into the studio and had Tom Barker set up a camera on me and said, "Just have conversation with me." So we did. I had two or three more meetings with him and finally he offered me the job. This was May 4, 1984. (I found out later that the rest of the board of directors was opposed to hiring someone who had not been in TV for so long and had no impressive credentials in another market in weather). But like the chick just out of the egg…I was "born" and my first reaction was "now what?!" I was put under the wing of Adrian Gibson for several weeks...so much of my early molding was thanks to him.

BEING MAGIDIZED!

I remember the first several weeks of trying to fill the shoes of bouncing Bob Baron! It didn't take long for me to figure out, you DON'T fill his shoes, you just try to blaze a new path. I had noticed the local anchors were dressing in mostly dark and drab, so I got the idea if I would wear loud and bright coats, it might stand out…I might get noticed as a new, if not really different, personality in the market. MD asked me once about those really loud colors, but decided, it couldn't hurt anything. It wasn't long and I was known as that dude that wears the green and other obnoxious colors. Our consultant, Magid, Inc., was also trying to figure me out and was continually offering their expert opinions (sorry, Steve Ridge). But, one thing everybody decided…I was becoming the "Mr. Rogers" of local weather. Therefore, I was sent to Iowa three times to get the "hello boys and girls, can you say pressure gradient" out of me. Once that was accomplished, they later started to work on my credibility as a weather authority. (That was a project that never ended, by the WAAY.)

HURRICANE COVERAGE HELPS BUILD CREDIBILITY

It wasn't long after being on board at 31 that MD and Magid decided I needed to travel to the coast to cover any significant tropical event that might affect us in our area. I remember my first trip to the Gulf to do live to tape reports (before satellite) on an approaching hurricane. Being my first experience, I showed up on the beachfront wearing my best and most expensive GREEN wool blazer….needless to say, with the sea spray and 80 mph winds, my coat was reduced to a size that might fit a Ken doll. I was too humiliated to say anything about it, so I ate that expense and learned a lesson. Nobody was doing reports from hurricanes at that time, so I had to learn the hard WAAY.

News Photographer Al Whittaker soon became my partner in all subsequent hurricane coverage. Al and I traveled to Florida, Alabama,

Louisiana, and Texas coastlines. In the later years, we fed reports back live via whatever satellite up-links we could buy, borrow or steal. I remember the hurricane "Gloria," heading directly toward North Carolina. We left in the wee hours of the day before it was to hit shore. On the way, we ran out of gas somewhere in the Georgia mountains. Al set off on foot to find help at 3 a.m., while I monitored the storm on weather radios. Al showed up with a state trooper and a can of gas about an hour later, meanwhile I'm learning the storm has taken a more northerly course. We plotted a map and decided it could hit Virginia Beach, so we changed course. Gloria actually turned entirely away from the eastern coastline and time we got to Virginia Beach, all we found to report on was the abundant supply of beach bikini wear that was bouncing all around us. I found out later that TV evangelist Jerry Falwell had been busy praying with his audience that the hurricane would turn away from the U.S. mainland. (His college and TV studios would have been in the original forecasted path.) I don't know if he had any influence, but I told Al, we'll never try to catch another hurricane that any TV preacher gets involved.

ICE STORM OF 1985

During most of my years at 31, we employed a weather consultant for forecasting. In February of '85, he was adamant we were about to have a big ice storm. Ice, as you know, is much more devastating to the public than snow. So, in one of my "out on a limb" offerings, I called for ice while everybody else was saying "a little snow." (Even Grady Reeves over at 19 admitted on air later that we were the only ones to get it right.) So, we got the ice and all power was shut down for Huntsville to the Shoals for several days. (Huntsville utilities got the station's power back on within 36 hours.) But my power at my house, next door to 31 studios, was out for four days and nights. Since I had no heating, my wife and I moved into the lobby of 31 during the period. That was the old lobby where huge portraits of the 31 anchors hung around the walls. I remember Ginger, (wife at that time), and I trying to get some sleep on the lobby couch. Ginger let me know that first night that she could not sleep with Heather Burns staring down at us. I had to take down the Heather portrait every night so we could get some sleep or stay warm (if you know what I mean).

TORNADO OF 1989

In the days before doppler radar, we had to rely on simple reflectivity of storms. While there was no tracking or even identification of rotation, we could see a strong line of storms moving toward us. That was the case of a line of storms coming out of Mississippi on the afternoon of November 15, 1989. We had been in wall-to-wall coverage since around 3 p.m. because of reports of damage in the western counties. I was on the weather wall at 4:36 when our producer, Ron Rutherford, handed me a note on the air that stated "a tornado on the ground at the old airport"…I stopped mid-sentence and looked at him and said…"a tornado on the ground"...and he starts offering my lead…"on the ground…at the old airport…moving toward Airport Road." It turned out Ron had been monitoring the police frequency, and the academy was at the old airport. They radioed to headquarters the fact of the tornado on the ground about to hit them. It turned out this was the only warning anyone had of the huge storm. The weather service put out a tornado warning four minutes later. By that time, the storm was almost to Gurley. As you know, 21 people died at Airport Road and on into Jones Valley, and many were horribly injured. My job was over, but our news department did a fantastic job of reporting rescue and recovery.

GARY SAID IT WOULD BE LIKE THIS

This was a promotional campaign in the never-ending chore of trying to build credibility for your humble weather-dude. But thanks to extensive newspaper full page buys in every major city in our ADI, billboards all across the area and bumper stickers galore...it was a tremendous success for the station. Years and years later, to this day, people still come up to me often and ask..."you say it was gonna be like this?"

Because of its success, some enterprising soul put another bumper sticker on the market that read. “Gary Lied.” I understand they sold out everywhere. I always accused Bob Baron (who was my competition at 48 during that period) of being responsible, but he denies any connection. (Yeah, right!)

WEATHER ON SITE

This was actually more an effort to bring 31 to the people. I did weathercasts from some really bizarre locations. Weather from the Botanical Gardens was a tie in to Gary’s Garden, a backyard working garden. But how about those weathercasts from roller coasters, from a seat dangling below a motorcycle on a high wire, or from the top of the nearly finished First Baptist Church steeple. When 31 hit the road with its newscasts, it usually fell my lot to be in the strangest spot or talking with the most outrageous guests. (Those were the times we but the credibility thing on the back burner.)

Okay, I could go on and on...but, I really want you to buy the book!

Let me sum up and say: I appreciate MD going out on his own limb to hire me in the first place, and then to support me thru periods when some folks wondered if he had lost his senses. I have fond memories of my 15+ years and miss so many folks I’ve come to love and appreciate along the WAAY.

"Cactus" (Robert A. Gay), Chief Engineer, Money Saver +

Worked from 1959 to 1999 longer than anyone else.

Dear MD,

As to me writing some of my thoughts of the past, I have so many I don't know where to start. So, if it is alright with you, you could start a folder for me and as I come up with some goodies you can save them...I remember so much about things that took place over the span of my life that I don't really know where to start...I also must be careful not to get my ol' ass sued by some of the "characters" that are still alive or would read some of this stuff.

(Censored - - -) She use to bring me and her a quart fruit jar full on frozen margaritas every morning and after the show we would either go out back or behind the transmitter and unwind a little...I remember this date very well, July 2nd 1963...I felt so good I decided to reward myself a little further so I went down to Electronics Wholesalers and talked Jack Pickens out of a new Collins S Line, which I still have today...

And the time the engineers, I think Bob and others, were transferring XXX movies from the film chain to VCR after sign off (turned main transmitter off) and they left the GE exciter on (which fed a small signal

through to the antenna) and everyone on the mountain witnessed the whole thing...

So MD, if you think we could have some fun with the lighter tales and a little later on some of our serious accomplishments such as our first Townsend transmitter which we paid $86,000.00 for and channel 19 paid over $300,000.00 to RCA at about the same time, the Double Cola case for the multiplexer for the old RCA film chain instead of paying RCA $34,000.00 for a new multiplexer and of course one of my favorite tales is when RCA and GE both wanted over $250,000.00 back in 60's to the sell us a color transmitter and I went down to Miss Mock's and spent a little over a dollar and got me some 47 millihenry chokes and a few trimmer capacitors to make some 3.58 MHZ color traps for the old GE exciter's modulator in order for it to pass the color burst and at the same time have back porch clamping...We use that same circuit until we retired the GE back in 1981...

Lew Koch–Operations, on Creativity, Picnics, "The Window"

THE SUMMONS: SHOWDOWN AT THE HOUSE

I arrived in Huntsville on July 4, 1979 and started working at WAAY-TV after the holiday weekend. Of course, I had interviewed with M.D. in June and found him to be friendly and easy to talk to. So, eagerly, I began planning a complete reorganization of the Production Department and was confident I could adapt systems successful in California to this new position. M.D. was very supportive, but cautioned me that his father, the C.E.O., would have final approval. I was a bit apprehensive because I'd heard Mr. Smith III was a tough manager from the "old school." Well, the man wasn't overjoyed at my radical proposals and summoned me to a meeting at The House. (What everyone called the corporate headquarters, the executive mansion.) M.D. accompanied me one sunny morning down Monte Sano, off Governor's Drive south onto DeRussey Road (as I recall) to The

House. We sat in a paneled meeting room and I was allowed to make my case for 15-20 minutes without interruption. Then the questions began. I couldn't tell if I was making any progress, but when it was all over, there was a moment of silence, then Mr. Smith III stuck out his hand and said, "Go ahead with your plans." That willingness to listen and change his mind made a great impression on me. I've never forgotten it and have tried to follow his example over the years.

BETTIE HIGGINS

Although M.D. made the decision to hire me and was gracious throughout the interview process, flying me out from California, I've always felt that that Bettie Higgins (Administrative Assistant at the time and, later, Program Director) was influential in my moving to Huntsville. We seemed to hit it off at first because we'd both been raised in Louisville, Ky. and we had a lot in common. e.g. she graduated from the rival high school, Manual. I went to Male. The annual Thanksgiving football slugfest was every bit as intense as Alabama /Auburn. Also, she knew about Herbie Koch, my uncle, and for years a star performer at WHAS, the historic clear-channel A.M. station. He was an organist (once employed by Paramount to perform around the country for silent movies). His radio, and later TV shows were popular (he beat out Pee Wee King for #1 local TV program one year). In fact, after being his guest at the studio one Sunday night I decided to switch my ambitions from journalism to broadcasting production.

One afternoon in mid-December 1980, when, obviously, I had time on my hands, tongue firmly in cheek, I sent the following memo to Bettie regarding a Production Department upgrade:

"What I need is total organizational capability in order to integrate reciprocal options. If we are to have functional management concepts, it stands to reason we also need compatible incremental projection. That is precisely why I'm recommending a balanced monitored contingency to give us the synchronized logic mobility Rocket City Television is known for. I'm sure you'll agree that a responsive digital time-phase using the new computer optical third-generation hardware will eventually give us the blop so necessary for success. Do you agree with this position?"

Bettie responded tersely: "Sure, same place as the last time. This time just remember not to use your credit card!"

And M.D. added: "Lew, I read your recent memo to Bettie. I was impressed. I, too, have been thinking of learning a foreign language...But then I wouldn't know what I was saying."

Ah, those were the good old days.

SNL—HUNTSVILLE STYLE: WAAY TOO LATE

I guess it was in the late 80's or early 90's that I was approached by several staff members with a radical idea. They proposed producing a weekly half-hour TV program with a comedy sketch format. It would be produced after sign-off, on their own time, and without any cost to the company.

Armed with a midnight time slot provided by our enthusiastic Program Director, Debbie Benson (Bradford), I convinced M.D. of the benefits to the company if he gave an OK to such an experimental project. We would have our own "Saturday Night Live." M.D., much to his credit, approved the project if I would act as executive producer (later joined by Debbie) to guard against objectionable material. Soon, an eager troupe of about eight regulars started production of the series.

Now I know I'm going to leave out some folks, but I recall those involved initially as Kip Cole, Lisa Waugh, Chris Homsley, Tom Barker, Jeff Rosado, Stan Williams, plus many others--a few from News--and special guests.

Although a few conservative department heads were concerned that bold satire might offend viewers and clients, the series developed a faithful late-night audience, served as a fantastic morale builder for the Production staff, and won numerous ad club awards. Not only did this series breathe new life into the regulars and guest participants, but the company also benefitted from new skills acquired at no cost. e.g. writing, shooting, acting, editing, set design, makeup, blocking, etc.

THE PICNICS:

At "The House"

Yes, those early summer picnics were staged at The House, the home of Mr. and Mrs. Smith III. It was informal, of course, but we were elegantly served a delicious lunch inside the mansion, and then we congregated around the pool or on the various patio levels to enjoy good food and conversation. One image I'll never forget is Tom Barker serenading Mr. Smith III and Cactus with his harmonica. There were always kids splashing in the pool and using the slide.

At Ditto Landing:

Bettie Higgins and I were partners on my first try at the egg toss. Guess what? We won! Other years I was not so lucky. The year Brent Smith won there were rumors he substituted a hardboiled egg, but Brent was a Production guy, so I knew he would never do such a thing!... And I'm still chuckling over the time I introduced two celebrities to each other. Now these picnics were great opportunities for me to impress new lady friends. One year, when the Bushes, George Sr. and Barb, were in the White House, my date was a brunette beauty named, you guessed it, Barbara Bush. Cactus was accompanied by the woman he later married, Joan Collins. So, with great glee, I introduced the two ladies, "Barbara Bush, I'd like you to meet Joan Collins!" Well, I thought it was funny…I recall Cliff Windom patiently calling out those bingo numbers. Most everyone played because the prizes were first-class…And some of you may remember that scene out of a Fellini film on a sunny afternoon, folks spread out on the green lawn playing games (sack racing, treasure hunt, tossing a football), when out on the water we discovered a river boat, a paddle-wheeler, enriching the picture.

THE WINDOW

I inherited M.D.'s old office when he moved to more spacious surroundings. He left behind a landscape he painted of a sailboat on a lake. It stayed on my wall for 20 years and is now in my home. (Yes, M.D. knows about it!) I don't know how M.D. managed all those years cooped up in an office without a window. I felt so confined, especially, in such beautiful surroundings outdoors. After a number of years passed, I decided to approach M.D. about this depressing state of affairs.

So I scheduled a meeting and pleaded my case. He listened quietly and didn't speak until I had finished. Then all he said was, "OK." I must have looked flabbergasted. "OK? That's it? OK? I've been putting off asking you for years and all you can say is OK?"

He smiled and suggested I find someone cheap to knock a hole in the wall. I did, a former clown, who most definitely knocked a big hole in the wall, then took off and was never seen again. I found someone else to finish the job.

Over the years, through that magic window I have observed bluebirds, dogs, snakes, rabbits, cats, etc. and, undoubtedly, my quality of work life improved.

THE JOYS OF MONTE SANO

What a great environment in which to drive to work every morning! Coming up the "back way" in spring with the sunroof open was a pleasant way to start the day; and if I drove down at sunset the southern route, or at dusk, glancing quickly to the right as the lights twinkled below, well, that was relaxing after a stressful day. I've seen coyote on the mountain and the rare Pileated woodpecker in the park; I've enjoyed many lunches at a park bench after a short hike, or on the lawn at the Burrett Museum. I've always said the WAAY-TV parking lot was the most difficult place to be when winter ice formed, not only underfoot or tire, but from above, from the towers.

My first day in Huntsville was on a holiday, July 4. I was alone, had been driving for three days and somewhat depressed by the Five Points area where I found myself. I drove to the station not knowing what to expect. For some reason, Elaine Smith was at the front desk that day and greeted me with a smile. She lived with her daughter on the mountain in a small cabin at the entrance to Monte Sano State Park. She recognized the state I was in and suggested I rent a cabin in the park and just wind down over the weekend. Some of the best advice I've ever received. Elaine became a good friend over the years and I am still saddened by her untimely death in Kentucky.

SENIOR SWINGER: JOHNNY EVANS

During a long stretch in '92, WAAY-TV produced a country music program called "A Country Affair." It was hosted and produced by Johnny Evans and directed on tape by Tracy Slayton on location at The Hop, a local club. As I recall, James Hessler worked the roving handheld camera, and I was executive producer of the show. This was quite a production with a live audience sitting at tables enjoying the music and dancing out in front of the stage. Between acts, Johnny would interview the performers while sitting behind a small table covered by what appeared to be a skinned cowhide (black splotches on white). Johnny brought in some famous country acts (though somewhat past their prime), and some less famous which I'll get to in a moment. On stage during the series were Little Jimmy Dickens, Patsy Montana, William Lee Golden, and Mr. Dobro himself, Tom Swatzell, who died late last year.

I'll never forget the March 15 show because my folk trio, "Legends of The Past," sang two numbers. This was only our second year (out of a 10-year run) together, and I feel we did OK, but the kicker was our interview with Johnny. I still get a laugh every time I views the tape. After introducing and briefly interviewing each of us, Johnny proceeded to give a lecture about the state of country music. For example, he spoke of "music blending together," and that "rock and roll of the 60's and 70's is the country music of today," and that he could "forego heavy metal and some opera--though not Grand Old Opry!" It was a classic discourse. Eventually, he got back to the three of us sitting beside him. "Is there anything in particular you want to talk about?" he asked me, winding up the interview. I just smiled and said, "We work cheap."

TIDBITS

Remember that long, wooden stairway that M.D. had built down to the lower parking lot? For a time, I was the only one who parked down there so I would be forced to exercise building up my wind and leg muscles (I played a lot of tennis.). I used to know the exact number of planks on the stairs, and if I found repair necessary, I would report the number to Junior Morris who would take care of the problem. To encourage more folks to use the lower parking lot and relieve the congestion outside the station, M.D. would offer prizes in a drawing for those who parked down there. . .

WAAY-TV offered many opportunities to women in television. One of my favorite courses in grad school was Women in Film, so I was especially interested in seeing equal opportunity. Now I KNOW I'm going to leave out a lot of super staff members, but to name a few outstanding achievers in this category, there was LIZ WEBB, the first lady switcher at 31; Location Supervisors (photog/editor, etc.) DEBRA NAPOLITANO & KATRINA BOVEY; Art Directors LEE BAKER & KYLE JESS; Operations Coordinators JAN MILLER (I still have video copies of PSA spots she and I appeared in) & LINDA POPE; creatives such as LISA WAUGH, so instrumental in producing the WAAY TOO LATE shows (and, privately, she directed my one and only music video in a downtown graffiti-alley on a cold fall day); Commercial Producers, for example, the multi-talented PRISCILLA COWING, etc. Those are just a FEW names of women at 31 who also staffed studio cameras, weather graphics, audio, and chyron. . .

I would like to spend some time on Commercial Producers and Newscast Directors, especially the GUYS, but I gotta wind this down. . . . Here's a final tidbit. I once hired a commercial producer out of South Carolina named Kelly Cauthen. His long-range ambition was to attend the American Film Institute in L.A. and, by gosh, one day, that's just what he did. A few years later, I was reading at home, the TV set was on, tuned to Showtime. I happened to look up and noticed the credits read "Kelly Cauthen, editor." The program was not family-oriented, but an R-rated you know what. The phone rang. It was Priscilla. "Did you see that?" she asked. "Kelly has a knack for body parts!" (I'm paraphrasing here.) After hanging up I was amused that Priscilla thought I would be watching that particular program! It wasn't long afterward that I "accidently" caught Kelly's name as "Director," then "Producer." . . .

Quite a long list of former Production staffers have done very well in the outside world and the list goes on and on and on…I've left off stories about the Addys, the telethons, the E.O.M's, the Fall Color Cruise…Maybe another time.

So long. lk May 18, 2003

Dan Akens - Early ND, Studio Disasters, Fake Fire, & Paradise

[7 1/2 years at Radio & TV) Stories: WAAY of the World, Tiger and the Go-cart, Atlanta on Fire, Christmas parade.]

Photo of Dan and other is Dan and Val Ginter playing on 31 basketball team in 1966.

Dan Akens was hired as WAAY-TV's first News Director/Anchor, then left to run the TV Studio at Marshall Space Flight Center, later during 12 years at WSMV in Nashville. (Won a Peabody there and numerous Emmys) - Most lately General Manager of the ABC Affiliate in Tallahassee, Florida for 8 years. Currently on assignment for the same company at KADY-TV in Ventura County, outside of Los Angeles.

#1...The Fish Tank...WAAY-TV's first newscast was called..."THE WAAY OF THE WORLD"...A title first coined by MD III and used extensively by the radio station. The graphics for this first newscast were done by rear screen projectors and the Anchors stood up at a podium, with hopefully just enough light on the news reader, so that you could still see the background graphic on a screen the size of Wyoming...In the studio, we had also begun a morning studio talk show called..."COFFEE BREAK"...This had the traditional couches and chairs, and one "homey" attraction...A FISH TANK...

In the early days, the first News Director/Anchor...Dan Akens, also shot, processed and edited film, and somewhere in between wrote

copy...One night when arriving on the set while the "WAAY of the WORLD" theme was playing, Dan discovered that the Fish Tank, which was about 50 Gallons of water and 50 guppies and assorted tropical fish, on a wrought iron stand...was right up next to the podium, where he needed to stand, (to catch that aforementioned light)...He put the script down and tried to "schootch" the tank out of the way...

The tank of course went crashing to the floor, with glass, water and fish flying everywhere...Soaking wet from the knees down, with fish flopping and cutting themselves on the broken glass...The floor director cues him...(O.K. No Floor Director...The Studio Cameraman cues him)..."Good Evening...from Huntsville...This is ...The WAAY of the World!!!!" (When the first film roles Dan tries to pick up as many flopping fish as possible and put them back in what's left of the broken tank.)

#2..."Tiger and the Go-Cart"..."MD" thought we needed an afternoon children's show...(Not having learned his lesson with Benny Carl) We took a DJ from WAAY Radio...named Tiger Terrell, put a Pith Helmet on him and thus the "TIGER TERRILL SHOW" was launched...There was only one problem with Tiger Terrell hosting a children's show... Tiger was still a CHILD, himself...One day a sponsor had a Go Cart in the studio...During a station break, Tiger decided to start the Go-Cart...IN the studio and began to "spin" around the studio crew, going "Whee!"...until he

came in contact with the pedestal of Camera Three. (We only had two cameras but it sounded better to call one of them "3".)...The camera and pedestal shot to the concrete floor, smashing the lens turret...and basically giving "Cactus" a pretty good sized project...(Fortunately, all the Alumni know "Cactus" Gay...and I don't have to explain any political correctness in regard to his name...he was born somewhere in the prop storage area of the old studio, and just never left.)

Well, we did the rest of the "Tiger Terrell Show" AND the news with only one camera, while MD wondered what was holding up Tiger's Pith Helmut?????

#3..."Fire in Atlanta"... Dan Akens had worked for Smith Broadcasting for four years, first at WAAY Radio, Then at WJIG in Tullahoma, Tn., at WNUE in Fort Walton, and finally with MD at a production studio for all the stations in the Smith Broadcasting Offices in Homewood, Alabama. Dan left to go do news, as Ex. Producer of the Georgia News Network at flagship station WGST in Atlanta. While working in the newsroom one day, the switchboard called Dan and said he had a phone call from someone who said, there was a fire at the Heart of Atlanta Motel and they could give him an eyewitness report...When Dan got on the phone, it was M.D. Smith, III, trying to disguise his voice (which was impossible...he could do several different voices, but if you knew M.D...you knew all the voices were his) Anyway, after saying..."there's fire at the Heart of Atlanta Motel"...Dan said..."What are you doing M.D.????"...M.D. said..."we've just bought the TV station in Huntsville, and I want to talk to you about coming back to work for us as News Director/Anchor,"...And the rest is history! (Here's Dan with a 16mm Bolex, the staple of the TV news industry for silent cameras.)

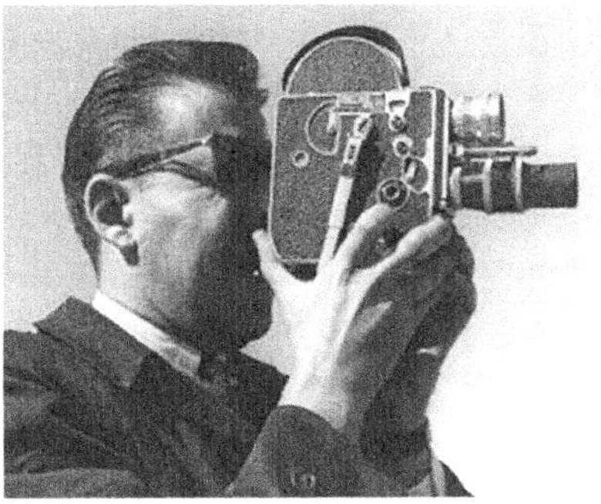

M.D. IV...Stop me, before I hurt myself!!!!... Here's one from WAAY Radio days, but since you guys adopted the Christmas Parade, you may not know this story...

#4..."Lights and Sirens"... This is about the first WAAY RADIO Christmas Parade...For whatever reason, the city would not give us a fire truck for Santa to Ride on...Maybe because they had their own

Christmas Parade...I can't remember...But I know that we put on the first WAAY Christmas Parade and it was to end up at Parkway City, which of course sponsored, the whole thing...We read Christmas Stories on the air for ten days leading up to Santa's Big Day...Miss "Merry Christmas" read the stories (Many of which I wrote, because there weren't that many good stories then and MD wanted 10 days of advertising out of this)…and then with MD supplying the sound effect of the short wave radio, we would contact Santa and see where he was on his trip...but I digress...We couldn't get the fire truck, so we decided to use the convertible (kind of maroon, but almost red) of "Wild Bill Malone"...the night-time dee-jay at WAAY Radio...He was the Mayor of "Lick Skillet," up there near Fayetteville...

Well, there we were in the WAAY Christmas Parade, Miss Merry Christmas in her little red outfit on the back of the convertible "Wild Bill" behind the wheel and I was in the passenger seat, when I discovered that under the dashboard, Bill had some switches for flashing lights inside his grill work and a siren that sounded just like an emergency vehicle…Sooo, I laid on those numerous times during the parade and kids loved it...(Did I say Tiger Terrell was a kid?) ...Anyhow, when we pulled up at Parkway City, two cop cars asked us to pull up by the front of the mall and they began to look under the hood for the "lights and sirens"...Seems they had been looking for a car that would flash and run its siren, up and down the Fayetteville Highway !!!!...I think M.D. got "Wild Bill" off the hook, paid the fine, or whatever...All for now...Sorry to get carried away.

Dan

Bill Powers' Story—April 1974 Tornado Outbreak & Details

April 3rd, 1974 is a date I'll always remember from my years in Huntsville and especially those at WAAY.

The day had passed uneventfully - as had "31 News at 6" with the typical assortment of national and local stories, including the usual accident film and voice-over from Dick Van Valkenburg - followed by weather, sports, and a wrap-up. As best I recall, the team on-air that night included Adrian Gibson, Tony Beason, John Bradshaw, and Dan Jamison. *[Editor's note: Sam Depino, Adrian and John Bradshaw did the bulk of on-air work. Bradshaw using our new radar (with weather service help) pointing out the "hook echo" and Sam warning public to take shelter and Adrian in another part of studio.]* Larry McCree and I were in the control room.

Our new "toy" at the time was an RCA AVQ-15 weather radar. Cactus, Bob Sullivan, and Don Roden had worked feverishly to get our radar on the air ahead of WHNT who had sawed off their 150' tower to install a HUGE Navy surplus radar (we joked about WHNT being able to warn people of ships coming up the Big Spring). Neither system was originally designed for weather duty. Our AVQ-15 was an airborne radar adapted for ground use. It was hard to see much through the ground clutter on the display, but properly "tweaked" it worked pretty well. We made our goal of being first in the market to have one, and it paid for itself the night of April 3rd.

Earlier in the afternoon a tornado had been reported just west of Birmingham, followed by another one east of the city about an hour later. We didn't think much of 'em and they were at best out on the fringe of our viewing area.

At 6:30 p.m. we switched to "NBC Nightly News" and had begun the routine of shutting down the studio lights, capping the cameras, and pulling the slides and news film for use on a "31 News at 10" that would never come. Hardly five minutes into network news a tornado warning for Lawrence County came over the teletype from the National Weather Service and we were back on the air - with John Bradshaw - white as a ghost - declaring the emergency live to our viewing audience. In a matter of minutes, we identified what we believed to be a "hook" pattern on the edge of the radar screen's 20-mile range - indicating a large tornado somewhere out there just beyond Decatur and which seemed to be moving our way. This was followed by another tornado warning at around 7:00 and another sighting on our radar.

The bright skies of only an hour before had turned ominously black as large hail and heavy rain pummeled the building. More tornado warnings followed as the line of intense thunderstorms continued to build over western Morgan and Madison counties. Sometime during the evening, the National Weather Service station at the airport was abandoned. Instantly, we became the only operating weather radar between Centerville and Nashville and Memphis and Atlanta. I'm not sure which was greater - our apprehension over the screaming winds around us or the excitement of knowing that at that moment we "owned" the ratings.

Harrowing though it was, it was punctuated by moments that will "live in infamy" - at least for those of us that were there.

I remember while John was on the air at one point Larry saying, "Look out, he's in trouble!" We had five "hook" patterns on the radar screen and John was hyperventilating! He couldn't catch his breath. We cut away to the radar screen and killed his mike while someone handed him a paper bag to breathe into. I can't remember whether we were shocked or just laughing or asses off (I think the latter).

The storms were so severe that electrical power eventually went off across most of Northern Alabama - taking us off the air just before a tornado struck Monte Sano. *[Editor's note: This was 10:55 p.m. that night.]* I remember looking off to the west out the back door of the station and seeing several funnel clouds - glowing green from highly charged ionized gases - coming our way in the vast darkness where Huntsville and the horizon should have been.

We thought we were safe on the mountain - the general consensus being that updrafts would pick a tornado up and carry it beyond us. But around 11:00 p.m. we knew we were wrong.

I'd never been in a tornado before, but there was no doubt about what was going on outside when we heard the roaring "locomotive" sound we'd been told of so many times. In an act of immense journalistic bravery (a.k.a. insanity) Tony Beason - cassette recorder under his arm - held a microphone out the double glass doors at the front of the building to "capture the sound of the tornado." No doubt he envisioned how dramatic it would be on the air the next day. I, on the other hand, imagined those glass doors shattering at the moment the building imploded. It went along with my vision of being crushed under our 500-foot tower or channel 19's 1000-footer - or both.

Bobby Gleason lived in a small stone cottage near the southwest intersection of Monte Sano Blvd. and Panorama Dr. I remember him walking into the station about an hour after the power went off. He had a cut or two and had obviously been through a rough time. He related to us how he looked out the window of his house to see his girlfriend's car disappear - vertically. I don't think they ever found it. The home next door belonging to Huntsville police sergeant Bob Owen was completely lifted off its foundation and moved a few feet.

Though the station escaped any serious damage - it was pretty obvious that we were going to have a rough time getting past downed trees and other storm damage. As time went by we would all learn how much more serious the day's events had been. Martial law was declared in much of Huntsville to prevent looting. We had a member of the National Guard watching our street for a day or two.

My wife, Susan, and infant son, Bryan had been safely tucked inside a neighbor's storm shelter along with WAAY Creative Services Director, Hank Price, and his wife, Maria. My house was only missing a shingle or two, but Hank's house just next door had a collapsed front porch. Only a block or two away the damage was much more severe - bare foundations where homes had once been. And there were the oddities one finds after such events - like the house on California Blvd. where nothing was left standing but an un-touched bookcase (filled with books).

Fourteen states had seen the brunt of this intense weather system. Later studies would conclude the system produced at least 148 tornadoes, 30 of which were classified as F5s (260 MPH+). One F5 had blown away Guin, killing nearly two dozen people there. The towns of Harvest, Hazel Green, and Tanner were also seriously damaged.

Of course, the heroes of the storms were the many police, fire, and rescue workers who responded to those in need across all of North Alabama. However, those of us working at WAAY the night of April 3rd and those in the field and reporting over the following days couldn't help but puff our chests out a bit as we relished a behind-the-scenes panic viewers never and as we contemplated the service our station had provided the community. Kids in the business today have no idea...

The Main Studio and 31 News set before the NBC News Center was constructed (that was replaced when we switched to ABC in December of 1977).

More from Bill Powers: The mind is a funny thing. A few nights ago, I dreamed I was in a plane that had successfully crash-landed someplace. Trying to summon a rescue I kept calling, "KGI-319, this is KFG-925." (The VHF two-way call signs between the station and whatever news crew was out there.) Nowadays I'm given to forgetting important stuff, but I remember those two call signs. Go figure.

I've attached a photo from the early 70s. Television was fun back then and there was a lot of it to go around at WAAY:

The Santa Show (w/Hank Price as Santa). The show closed with dissolves between cameras panning the shelves of toys. At the end of one show the camera panned across to find Ken and Barbie dolls that one of our resident funny men had placed in the "missionary" position - Ken's pants around his ankles and Barbie's skirt hiked up around her waist. I didn't catch it until the camera operator (Mike Howard, I think) panned back to see if that was what he'd actually seen...live!...on air! Fade-to-black! Go to commercial! Wait for the red phone in the booth to ring. (Fortunately, I don't think we got any complaints. Either they missed it or the audience size wasn't quite what we thought it was!)

Walton Jones had a habit of coming onto the weather set only seconds before we came out of the commercial break - often leaving us no time for a mike check. One night, fun-loving Bob Trotter arranged a loop in the microphone cable right where he knew Wally would be standing. True to form, Wally walked onto the set only about 5 seconds before the beginning of the weather segment. Just as the tally light came on, Bob - standing just out of camera range - pulled the slack out of the cable, tightening the loop around Wally's ankle. For the entire 3 or 4 minute weather segment Bob kept the line tight and Wally did the weather as a stooped-over guy dragging a bad leg. Thereafter he showed up on the set at least 30 seconds before he had to go on.

On occasion I've been asked, "What's the best way for my [insert relative] to break into the business." After I tell 'em most people are trying to break out I recommend they rent two movies - Weird Al Yankovic's "UHF" and "Broadcast News". If the kid still wants to be in television after those doses of reality - he should go find a small or middle-market station where it may still be possible to learn how things work.

I had many learning experiences at WAAY and survived at least four career-threatening events: The darkroom enlargement. Decided on my own to connect what had been the old film processor room and my closet of a darkroom adjacent to it. Good idea - badly executed. Didn't think about the dust from that cinder-block wall I took down being sucked into the return air vents - only to be deposited on most flat surfaces throughout the building.

My first encounter with Mrs. Smith, III. I was running master control during the Today Show. She walked into master control one day, introduced herself, and started up what seemed at first to be a friendly conversation. I was thrilled to meet her and made no secret of my appreciation for being at WAAY. Mrs. Smith asked me, "So...how long do you expect to stay employed here?" "A long time," I replied. "Well...young man...you see that sign? [Keep this door closed] You won't last until the end of the day if I come back here and find that door to the video tape room open again!" Needless to say, I got the message.

The night I walked through one of the double glass doors by Johnny and Redge's office. Bill Farris drove me down to Huntsville Hospital to get sewed up. Still have the scars. Only thing more painful was having to explain it to you the next morning.

The camera blow-up. Actually, the camera didn't, but Cactus did. I had just finished eating my lunch on the Coffee Break set while watching Cactus - standing on the camera pedestal - try to fish a screw out from deep inside one of the TK-42 studio cameras. I quietly blew up the bag - snuck up behind him...and BOOM! Damn, I thought we were going to have to peel Cac off the ceiling! And man, was he pissed! Fire-breathing - Pentecostal preacher-breathing in-my-face furious! I don't think he cooled down for at least a couple of days, but he left no doubt about his holding the life-or-death of my career in his hands. Thank goodness for the benevolence of the Smith family!

Val Ginter's Stories—"Bitch Box", Dark Shadows, 31 Funtime, and more

Golly, M.D., that's like opening the floodgates! We were the first Mary Tyler Moore Show. I tell WAAY-TV stories almost every day of the year...not to mention whenever somebody asks about the clown hanging on my bedroom wall from the Name the Clown Contest (it had been thrown into the GI can after the show and either Ernie Iler or Bobby Gord presented it to me as a souvenir).

One of the funniest, or unfunniest, was when Jerry Robinson used to hand Terrell Whitaker (Tiger Terrell) the football scores during the Community Bulletin Board show (sponsored by the local lumberyard). He'd hand Terrell a tear sheet from the machine and Terrell would yell: "Roll Tide! Hey, Alabama's creaming Purdue!" And then he'd hand him another one and Terrell would yell: "On Wisconsin, On Wisconsin! Smother those guys at Illinois!"

And then Jerry handed him another tear sheet and he yelled: "Hey, Jayne Mansfield! Va, Va, Voom! Va, Va, Voom! Hey, let's see what this says. Jayne Mansfield's son has just been killed in an automobile accident. Oh, my gosh. Oh, that's not very funny. Oh, that's awful. Oh, I'm sorry everybody. Golly, gee. Let's take a break." And then he looked at us in the studio during the commercial and said: "You guys are a bunch of bastards." And we were all doubled up. But we were in our twenties at the time. This would have been in the spring of 1966 or around then.

Val ********************************

Now you got me started, M.D.

When Sonny was Tony the Tiger on 31 Funtime, he used to come up into the control room during the theme song of Dark Shadows. You remember that eerie theme song (OooOooOooOooOoooooo). And Sonny would come up into the control room in his tiger suit and stare out the window to see the studio. I mean, he didn't want to go into the studio yet, because he wanted to make a grand entrance for all the kids in the gallery and the mothers sitting on the side. And he would be really nervous and laughing and giggling because he'd be going on in maybe sixty seconds. And he used to double up and laugh and say: "Boy do I love that theme song. Listen, listen, here's my favorite part." And there was just that eerie music, but at the end of the theme, the music would fade down, and the announcer would say very quietly: "Dark Shadows is a Dan Curtis production."

And every afternoon Sonny would come up into the control room and laugh and giggle and say: "OK, now let's be real quiet now. We want to hear the best part. Listen, listen, Dark Shadows is a Dan Curtis production. Isn't that funny? Ha, ha, ha!" And we'd go: "Yeah, yeah, yeah." And he'd do this every day, and we'd pay him no mind after a while. Then one day, he comes up into the control room and does the whole routine: "Listen, listen, I love this part. Hey, be quiet, you have to listen to this. This is funny!" And the music plays and then fades down, and the announcer says in almost a whisper: **"Dark Shadows is a Val Ginter production."** And we all cracked up. He had taken the promo music and recorded his own theme, then he had the engineer roll the cart and go local immediately after the final ABC commercial position. The practical jokes that can be played in broadcasting...!

(It was one of those discarded Fidelipacks (audio, endless loop cartridge tape) that we wouldn't put any commercial material on, and I took it with me to New York years later. Unfortunately, it was in such bad condition, it simply fell apart in moving. I think I still have it stored away in a trunk, but the springs and pads are all out of whack. I tried to play it at CBS, but it just wouldn't work.)

Val ********************************

Thanks for the great picture of Johnny Evans! And also, last week's pix of Sully (Bob Sullivan). He looked mighty healthy in the picture. I left in 1968, so he lasted up into his forties, I guess. We used to bet that he'd never see thirty with all his smoking and drinking. But he was a strong guy. And he had a lot of good attitudes.

More from Val Ginter

I enjoyed reading the Word attachment that you wrote on purchasing Channel 31. That Bolex interested me. There is a nice shot that Jerry took of me with the Bolex as I shot that political—Jim Martin or Dick

Martin—I forgot his name but he was running against Lurlene Wallace I think. Anyhow, there is this shot of me in the studio on that big GPL Camera One and another of me shooting Martin's whistle stop at the Athens railroad station with the Bolex.

I shot the silent footage, and Jerry Robinson used the Auricon for SOF footage (the day before Thanksgiving, 1966--it snowed like hell). There's another shot somewhere of Jerry on that Auricon. He plugged it into one of the light sockets above the Athens railroad station platform and was standing in a foot of snow wearing earphones, holding the panning handle with one hand, and holding the microphone with the other. He wasn't wearing any boots, and he had continuous electrical current running through him. But you remember Jerry—it made him giggle.

Val sends some more goodies - May 22, 2003

Here's another great black and white story. In 1967, Jack's Hamburgers sponsored a Name the Clown Contest. We had this little clown doll hanging on the wall of the 31 Funtime set, and somebody out there in TV-land had perhaps two months to come up with a name for the clown. The winner would receive one-thousand Jack's Hamburgers.

Well, of course, it wasn't really one-thousand hamburgers; it was one-thousand hamburger coupons. This way you could give them away when you got tired of eating hamburgers.

But for the studio, there was this display on one of those big folding tables that featured one-thousand Jack's Hamburgers stacked pyramid-style on top of the table. It was kept behind the curtain against the wall, and when it was time, the studio crew simply pulled the table out from behind the curtain.

Again, I must correct myself. The table didn't contain one-thousand hamburgers. It contained one-thousand empty stale buns wrapped in Jack's logo parchment paper. But it looked great on Camera #3 (which was a studio vidicon and made anything look good).

For two months, we did the Name the Clown Contest immediately coming out of the Three Stooges cartoon. Five days a week it was: "And now it's time for the Name the Clown Contest." And, originally, I'd go: "Three in tight on the hamburgers. One zoom in on the clown. Ready three. Take three. One on the clown and ready to zoom back to Johnny." I did this so many times that the cameramen began giving the cues in their mouthpieces, and I didn't have to say a word anymore.

Finally, after two months, some kid won the contest. The guy from Jack's came in and gave the kid his one-thousand coupons, and somebody from the studio crew ripped the clown off the wall and threw it into the GI can in the corner of the studio.

The best part of the story is when we took a final look at that hamburger bun display. Now those buns must have been stale to begin with; otherwise, they wouldn't have gone into that display. But talk about penicillin or microbiology or just, simply, mold: every hamburger bun was a different color. I became educated in science the moment I looked closely at those buns. Each and every one of those thousand buns was a different color: one was red; another was purple, another was green; another was blue; another was black; another was orange; another was yellow. We could have donated that table to the Smithsonian (or, maybe, Johns Hopkins).

Somebody--I don't remember who--suggested that if Bob Sullivan had extracted all the penicillin on that table, he would have been set up for life.

I think it was Ernie Iler who retrieved the clown doll from the garbage can right before the sanitation truck arrived and presented it to me as a sort of trophy. I've had that clown hanging on my bedroom wall for the past thirty-five years. Val

More from Val:

After the WAAY-TV reception room was remodeled and redecorated in 1966, the men's room received two doors, one from the reception room, and one from the studio, so that during off hours, the reception room could be closed to traffic and still afford use of the men's room.

Needless to say, throughout the working day, the men's room was used as a passageway by the male gender to get from one side of the building to the other without having to go outside into the elements. It was not a problem that you had to step over whomever was sitting on the toilet. It's all part of family living (and if you'd ever been in the army...) I guess.

Dan Akens, Bobby Hammers, and Dave Walker always saw to it I could get the Early News script by 5:00 p.m., so I could block it out. So, between 5:00 and 5:30 p.m., Monday through Friday, I would block out the script while sitting on the toilet. It took care of my constitutional needs, as well as providing me a quiet place to write in my camera cues, audio positions, and rear-screen pictures.

On this particular day, it was raining cats and dogs outside. And a few of the guys climbed over me as I sat on the commode. Then the door opened and Dan Akens walked passed me, and said "Hi Val;" then this dude, around fifty years old in an extremely expensive suit (a client, I figured) steps over me and looks at me rather strangely; he is followed by a very well-coiffured and suited woman of around forty years old, who looks at me even more strangely; and she is followed by a teen-age girl with long blond hair, dressed in a preppy outfit like the wealthy kids wear who go to school in England, who looks at me with an expression of extreme shock on her face. The last person in this parade was our salesman, Marvin Blitz, who walks past me, lets the other people out of the bathroom, and then turns around at me and exclaims: "Val Ginter! Have you no sense of decency?!"

I stayed away from the front-office area for the remainder of the day.

[Editor's (M.D.S.) Note: Val, thanks for some very good stories and memories. I had forgotten that at one time the men's room (which now has a hallway from "old" lobby to glass door entrance to building) had that door right in front of the toilet and that people used it as a passageway. Later, we built a wall and you enter the men's room between the toilet and sink, but you do hit your knees on the wall when you sit down on the toilet. There was not a lot of space left after we made the hallway.]

THIRD POSITION

After a few years of doing the prime-time routine, I was pretty well attuned to the habits of network positions during ABCs schedule. The mid-station break came after the third position of an hour-long show and ran anywhere between twenty minutes after and twenty minutes before. There had never been an exception.

Big Valley was probably the only program I didn't watch when I was on the board. It simply didn't appeal to me—all that nineteenth-century costume drama stuff. So, I loaded up four slide carousels, three film projectors, and our trustworthy Ampex VR-1000 for the next two or three breaks, and at around 9:08 p.m., I headed for the bathroom, which was up at the front of the building.

There was a "bitch box" standing alongside the bathroom door, so we would never have any problem making our cues. I was sitting on the toilet, concentrating, and doing whatever else one does at a time like that, when Big Valley ran a commercial position. I looked at my wristwatch (which was always set to the network clock beep) and saw that it was 9:16 p.m. So, I continued to do what I did best, sitting on the commode. At 9:17, the announcer comes on and, it seemed like at the top of his voice, says: "Big Valley will return after station identification." And then, there was the customary five-second bumper cue (in those days, everything was cued out five seconds): "See the Milton Berle Show, nine p.m. eastern, eight p.m. central, tomorrow night on A-B-C!" And then there was complete silence.

What could I possibly do? I took care of my needs as fast as humanly possible and grabbed my pants and ran across the studio with my pants still somewhere between my ankles and my knees. The studio separated the bathroom from master control, and we're talking a good one-hundred feet. As I passed the monitor in the studio, I see a big sign on the screen: ATTENTION AFFILIATES. And this guy with an Irish falsetto voice and a Brooklyn accent begins to scream: "Attention affiliates, attention affiliates! This is ABC control, Noo Yawk! This is ABC control, Noo Yawk! There will be a one-minute-and-thirty-second news story on President Lyndon Johnson's trip beginning at 11:05 eastern, 10:05 central, and preceded by a one-minute, thirty-second, fifteen-second, and five-second video countdown Thank you! This is ABC control Noo Yawk!"

And at that point, I was behind the board. Well, about all I could do was go local and roll the break. I believe it was a twenty-second Preparation-H with a twelve-second promo-id (the latter of which I cut). I never went to the bathroom again during prime time. (You'll find this somewhere in the discrepancy reports from 1966.)

Don Roden—Engineer & First Snowy Impressions 1968

As we approached Hampton Cove, it was getting difficult to see. The windshield wipers were struggling to clear the snow off the Ford Mustang's windshield. We stopped at a service station at the foot of Monte Sano Mountain. We pulled behind the building where the pavement was protected from the heavy snowfall and put snow chains on the Mustang.

"We will go to Lewters Hardware and get you a set of chains," the driver told me. But he cautioned, "If you lose them, you will have to buy the next set yourself."

As we turned off Hwy 431 onto Monte Sano Blvd, the snow was already a couple of inches deep. "How do you know where you are going?" I asked. "Follow the power lines." the driver said. I was wondering to myself, "What have you got yourself into now?"

When we pulled into the TV station parking lot, there were three cars outside the building.

Seeing no tracks in the snow, I asked, "How did they get here?"

"That's the receptionist, the production manager, and the owner's cars. They all live up here on the mountain."

"Let's go in and I will introduce you to them and to the most addictive job you will ever know."

And that was my introduction to WAAY-TV on that snowy January 3, 1968.

Cactus was the Mustang driver and M.D. owned one of the cars parked outside WAAY- TV.

Don Roden

. . . and now, the rest of the story. In the photo below, from that actual day, you can see Cactus in the driver's seat and Don is riding "shotgun" as Cactus said.

Santa's TV Electric Chair – 1963 – Unforgettable

By M.D. Smith, IV

It was the first Christmas in 1963 after we bought the station, and M.D. Smith, III decided we needed to have a Santa Show on TV. He got a salesman to sell the sponsorship to Montgomery Wards. It was decided to pitch their toys from Santa's "workshop" just to the side of the set. We all agreed it was not right for Santa to "hawk the toys for Wards," so we decided on Santa's Elf Helper, Miss Merry Christmas to guide kids on and off his lap and do the commercial breaks for the toys.

But of course, we needed a big Santa chair for the Jolly Old Elf to sit in. Remember, these were the days of ALL black & white TV, and our studio was painted shades of gray, as was all our artwork. Howard Troutman (artist when we bought the station) was very good at hand painting 11x14 art boards in shades of white, gray and black on a sort of medium gray board. He was a bit of a handyman as well, helping construct the sets. (With 17 people total, we all had to do a lot of jobs.) So, Howard set about building the chair out of 3/4" plywood so it would be strong.

Howard basically built a strong box with straight, squared angle arms on each side and a square, rising back of the chair that came about as high as a person's head when seated. Then he painted it a light gray that would balance with the other the shades of gray on the Santa Claus set.

It was pretty plain, I thought, but most would be hidden when Santa was seated in it, so I didn't have any comment about it. Howard Troutman was very sensitive about criticism on his work. However, when the Merchandising and Advertising Manager for Wards came up to the studio to set up the toy racks, he took one look at that plain, square, gray chair and said, " HOLY S%$#!!! YOU'VE MADE A G#$ DAMN ELECTRIC CHAIR!!!

"What are we gonna do, ELECTROCUTE SANTA, BROUGHT TO YOU BY MONTGOMERY WARDS?"

Well, he was right, now that someone said it out loud. It did look exactly like an electric chair, missing the wires and top cap. He had an absolute fit! I know Cactus remembers this well. Mr. Smith, III saw it and clearly agreed it was awful (but it didn't look all that bad on camera.) Mr. Smith, III talked to Howard (who got his feelings hurt very easily if you didn't like his work) and explained he wanted a big show for the moms and little kids coming to the station, and he wanted a THRONE of red velvet.

Time was short with the show to start in a day or two, so we first got another piece of plywood and made a large, rear, flower petal back for the "chair" and secured it. Then, we got a bolt of Red Velveteen and some sponge rubber and some brass tacks. We covered the chair and with the sponge rubber beneath the arms and back and holding it in place with the brass tacks, we actually did produce a very good-looking Santa Chair. The Montgomery Ward's guy was very pleased with the improvement. (He was actually a horse's ass to please and we had many more arguments with him that season.) The Santa Chair lasted for many, many years, but I bet only the oldest of Channel 31 alumni ever knew the real story behind the Santa Chair until now.

Carl Spurlock & the Santa chair/show follow up

[After reading about how the Santa Chair was made, Carl said . . .]

That explains many things, the least of which is why any of us (both you and I included) that ever sat on that thing while doing a stint as "Santa"...are not in traction for the rest of our lives! Damn, that chair was the most boxy, uncomfortable thing I have ever plopped my bum on! Add a chubby kid to your lap...and it gave new meaning to the phrase "pain in the ass."

I have one short Santa Show related story:

Being a classically trained actor, a member of SAG, AFTRA and DGA, I had always prided myself on never breaking character or breaking up on while on stage or on camera. That pride and my flawless record both came crashing down one afternoon in the 31 studios. Hank Price (production manager at the time) was carrying the burden of playing the "jolly old elf" this particular year. He had just finished the show and threw the commercial break. At the other end of the studio, I was making ready to come on, out of the break, with the first "Dialing for Dollars" insert as the afternoon block rolled into Star Trek. I was standing on the "D for D" set awaiting the count from the cameraman.

Hank, thoroughly exhausted from dragging one kid after another across his lap, came lumbering over toward the D for D set...pulled his beard down below his chin and lit a cigarette. We made eye contact for just a second and then I re-focused on "coming on" as the countdown was given. 5,4,3,2,1 and the finger wave "you're on"...up came the tally light and I was live on the air. Now Hank, usually a very serious man and all business when it came to the station, had apparently been push a little over the elf edge by the burden of being so damn jolly for an entire half hour. Without warning...Hank stepped forward right beside the camera lens...smiled at me, turned around, bent over and dropped his pants.

For the very first time...I had been caught completely off guard. The thought (not to mention the actual visual) of being "mooned" by this sleazy looking, butt puffing Santa just nailed me...I began to chuckle and then just flat laugh out loud. Of course, the viewers could see none of this. Just me losing it on air. I managed to say "Folks, we'll be right back" and stepped off camera. John Stanners threw up the Star Trek slide for 15 or 20 seconds as I collected myself. Quickly bringing myself back in control...again, the quick countdown...and we were back on air with the first D for D call of the day. True story...Carl got mooned by Santa (Hank Price) live on the air and totally lost it. Damn you Hank...I have never trusted a mall Santa since.

Regards, Carl Spurlock

Linda Allen's Story of "Fire Farters" on the Scene

The great thing about working in a mid-sized market is that your owners give you a lot of "growing" space. WAAY TV was my first tv news home and I sure did a lot of growing and learning during my seven years there.

I have so many fond memories of my life in Huntsville and my friends at WAAY but no doubt the most memorable occurred during the early morning hours when I anchored WAAY's morning newscast with Adrian Gibson and Gus Hergert. We were quite a team and responsible for a lot of shenanigans. Among my favorites the 4th of July 6 a.m. bar-b-que we had on the air with "line drives" off Monte Sano Mountain.

Also, Adrian stepping on "taped together" prompter scripts (remember when you had to do that) and getting them stuck to his shoe during the newscast. It made for a bumpy read.

Probably the funniest moments involve typos. (Most of us were young and never did get enough sleep.) One happened on a December 7th. I had over-slept and didn't get the "where are you call" from then-producer Alicia Smith until 6:15. I threw on a blazer over my PJs...hopped in the car and headed up the hill. I walked on to the set out of breath and with :30 to spare. I had to read scripts cold and was doing ok until I got to..."December 7th, a day that will live in infantry." Not only did I read it that way...a certain anchor "in training" read it that way later that morning during an 8:25 cut-in. I wonder if she's still in the business.

My other favorite was breaking news that happened one morning. I got on the air with very little info...but my trustee producer Alicia was in my ear with what I needed to know. We were doing great telling viewers about the dangerous fire taking place...we had great live video thanks to "Live Eye Reporter" Kim Albright and engineer Harry Hatfield... I was doing great until what came out of my mouth was "fire farters are on the scene." I heard some laughing in the back...but held my own and didn't break. That was until my producer gave me the next bit of information... "It was a GAS fire." I lost it then. In fact, I was laughing so hard we had to go to break because I couldn't recover. I'm surprised I wasn't fired on the spot.

WAAY will always have a special place in my heart. I met and made some life-long friends there. Like most families, we were definitely dysfunctional, but I feel so blessed to have been a part of that place. The experience changed my life.

Linda Allen

(Still in the business and still reading typos)

# # # # # # # #

Dave Hargrove – The GRAPE, Writing on the Computer "wall"

Reporter/Anchor

One of the greatest collaborations between M.D. and information technology director Mark Derrick was "the Grape", short for Grapevine. WAAY-TV was always a leader in the use of computing technology for business and "the Grape" would later be commonly called a bulletin board or chat room. This was in the 80's when you had to have a federal security clearance to know anything about what would become the Internet.

Disgruntled employees could log in to the Grape (at the time we thought confidentially) and type out whatever happened to be angering or irritating us at the moment. Whether we were complaining about management, an incompetent director, a poorly charted camera, a change in the benefit plan or a hideous outfit worn on camera by a reporter (I think some of my ties made some postings) - it would make it to the Grape.

To M.D.'s credit, he only intervened when postings started to get nasty and hurtful. I recall some positive change taking place—I believe as a result of some of these comments. I know I bought more ties.

[From editor, M.D. Smith - Yes, the Grape saved the writing on the bathroom walls (and much subsequent repainting) and I felt gave people an outlet for frustration, anger or whatever. I was requested to end "The Grape" long before it actually came to an end. There was always going to be grapevine and gossip, so allowing it out was one way to prevent pressure buildups. Also, Mark Derrick and I could go in and edit out anything that popped up that was "cruel or ugly" and simply not let something like that stay on the page. There were even some very funny poems and stories that appeared on the grape. I still have a few of them in my archives today. It was a good and valid experiment and like most good things, a few "piss-ants" killed it.]

Dan Jamison (humorous) Self Bio of 4-26-73

[Editor's Note: Even though Dan Jamison has not been with us for a number of years, he is remembered for his wit and humor. Here is a submission he made to promotion when asked for a bio to be put in a promo as well as to publications. I have a copy of his actual typed, double spaced, all caps submission with hand written corrections. A shorter and edited copy actually got made from this longer script that Dan wrote. I think he'd enjoy being remembered with this crazy script.]

(**FICTION**) Dan Jamison was born out of wedlock in a log cabin, on October 9, 1944. His mother was a poor chicken farmer in downtown Baltimore, and his father was either the local Purina dealer or one of the hired hands. Dan was a dull child, and when he said his first words, "Da Da" at the age of 7, it was known immediately that the boy was destined to have a raucous mouth with a talent for supposition.

Dan worked on his act for three more years before he unveiled it to the first grade at St. Cecelia's. The notices were rave...except in the deportment department. But Dan continued there until the 7th grade. This his dad discovered Annapolis, which is also a part of Maryland, traded off two of his oldest children for a 1949 Plymouth truck with no body, and part interest in an Edsel dealership.

Dan continued to flourish at St. Mary's School in Annapolis and played basketball and football for two years. He was graduated in June 1963, and it was not until August 1963 that it was discovered that Dan hadn't attended one class at St. Mary's since early 1961.

Dan was off in college at that time, a community college with a name too long to mention, suffice to say, that basket weaving was not what Dan wanted, and thus he left to go to Washington, DC and study at the National Academy of Broadcasting, an impressively named school, that Dan found in the yellow pages. He went six months, graduated, was $1,000 (American) lighter in the bank account, but had a new career planned for him by his uncle; Uncle Sam, that is, and Dan was off to the wars in 1965, where he had to fight the French and German women to keep them from getting his candy and nylons.

Dan got a job at radio in March of 1968, where in Warrenton, Virginia, as the first white janitor for the station, he quickly rose to news and sports director, after undermining everyone else's position, and getting a little dirt on one of the owners. Dan did play-by-play for the high school basketball and football teams for the next four years...80 basketball games, 40 football games.

In November 1972, in a moment of drunken craze, M.D. Smith IV summoned Dan to Huntsville, Alabama, a suburb of Nashville, and requested Dan take over the Sports Department, that up to that point was not being handled in the best fashion. Dan doesn't handle it in the best fashion either, but now the team has someone to blame it on.

Tony Beason about Dick Van Valkenburg, Wreck Coverage

(Dick was our hard news crime & wreck reporter.)

Dick Van Valkenburg was one of the true characters I've met in this business...and that's saying a lot! While I knew he had a home and visited him there on more than one occasion in 1973, it seemed as though he lived in his news car 24-7. It was a car to behold! Radios, scanners, cameras, wires dangling, dancing and spiraling. I don't think the backseat had ever met a human derriere, being covered by mementos accumulated by the best damn hard newsman in the business. Dick was totally professional and when I was at WAAY, he already knew more about cops, robbers and the legal system than I would for years. By example, he taught me you had to be aggressive to get the story. But there was one time Dick was taken and we all went along for the ride.

It was otherwise a typical wreck. A car rear-ended a pickup one night with no injuries. It was the type of wreck Dick would check out and we would not have aired...but this one had a twist. Seems when Dick arrived on the scene, there were a bunch of goldfish flopping around on the road. Dick asked a cop where the fish came from and he said, "the guy driving the pickup had been to a wild party, swallowed the fish while drinking beer (this was during the time when goldfish swallowing at parties was not unheard of) and had thrown up after the wreck, barfing up the goldfish." Seemed plausible and the cop told Dick that with a straight face. Dick shot the goldfish and wreck, came back and said, "Boys, I've got a great story." We reported it as such and hear from the guy that night threatening a lawsuit.

He was actually a fisherman and was going to use the goldfish as bait! Seems the cop, like every policeman then, knew Dick and thought he would immediately pick up on his joke. Another lesson learned, know the true source of goldfish before reporting!

Tony Beason

Erin Dacy's Unpredictable Cat Story from about 1990

One time on the noon newscast I was interviewing a woman from animal services who wanted our viewers to adopt animals. She was holding a gold and white cat. As our interview went along, I guess the cat got bored and tired of being held. He pawed across the woman's face as she was talking.

She was taken aback but kept talking about how nice and sweet pets are to have. She was bleeding a bit, and I guess felt it, so she kept moving the cat up closer to her face to cover up the bleeding scratch.

When she wouldn't put the cat down, he finally grabbed her by the jaw and bit her on the nose. We promptly ended the interview, but I bet that woman has never forgotten that TV experience. I sure haven't!

Rick Davis' Olympic Story – A True DISASTER On-The-Air

February 1980, we're in the middle of the Winter Olympics. It's Friday, just after we've finished the 6 p.m. cast and I'm on my way to Huntsville High for a high school basketball game. While there, word spreads that the U.S. Olympic hockey team has just upset the Russians 4-3. People immediately start to leave, because Channel 31 is carrying that game on a tape-delay basis in prime time.

Well, I'm back at the station, editing tape, and we're doing live teases from the newsroom. The co-anchor that night is Luanne (spelling?) Kahn. The hockey game is in the third period, the Russians leading 3-2, and ABC takes a commercial break. In the game track, we're about 2 minutes away from the tying goal, and then another 8 minutes away from Mike Eruzione's game-winner. Luanne looks at me and says, "Anything you want to tease in sports?" And I say, "Well yeah (you idiot, I thought), tease the hockey game, the U.S. and the Russians."

A few seconds goes by. The photographer says, “Stand by.” Then he points at Luanne. The first 5 to 10 seconds are uneventful, and then, almost as if in slow motion, Luanne says, “and in the Olympics, the U.S. upsets the Russians, 4-3.” I just stared at her. Later, people told me my mouth dropped open, then my eyes narrowed. They began grabbing sharp objects to keep them away from me.

It seemed like a day or two, but it had to be only seconds, and I said, “Luanne! What are you doing?!? Do you know what you just did??”

Too late. The phones in the newsroom had the old push-buttons at the bottom of the hand-set, and there was room for four or five lines. For 45 minutes straight, all five lines were blinking. Calls ranged from “You stoopid idiot,” to viewers in tears who said, “I can’t believe you ruined this night for us.”

The moral, of course, say exactly what you mean. I should have said, “Tease the fact that we’ll have highlights of the U.S.-Russian hockey game, but for God’s sake, don’t tell them who won the game.”

Anyway, that was an interesting night.

Tracy Slayton & A Very Strange Smell

During a commercial shoot, I used my tackle box as a prop. I then placed the box in my office where it sat for about six months. I started smelling something, and figured out that it was an open, yet re-sealed bag of catfish bait. (flour mixed with beef blood in small round balls.)

I got the idea that this would be a funny joke to place some of these bait balls in Bob Dowd and Jim Clark's office desks. Imagine the smell when they opened their desk drawers the next day I thought.

I thought wrong. The next day (I was off work) the entire front of the building from sales, to traffic, to Lew's office (My boss!!!) to the front lobby to the studio smelled absolutely rancid. To make it worse, everyone thought it was dog poop. I got called in on my day off to assist in cleanup, which I considered just fine, at least I was still employed!

The funniest thing about it was everyone blamed Shawn.

Tracy Slayton Commercial Production Manager

WAAY-TV Employed 1990-Present

Ben Boles – '94 Election Coverage & A Joke

Ben Boles, wife Erin Dacy Boles and daughter Ansley (born 5-20-02)

Before I came to work at Channel 31, I worked at WDBB in Tuscaloosa - and got the two stations to share election night newsgathering costs in 1994. So as Fob James defeated Jim Folsom on the heels of landing Mercedes in Tuscaloosa, our Tuscaloosa crew went to Montgomery to work with the "big station" in Huntsville.

Well the big station had two big people - Cliff Hill and Jeff Raker covering the story. I did not know Cliff, but everything you hope a good news man would be - he was. To those who worked with him in the newsroom - you knew Cliff never did anything quietly - both in an investigative sense and when he was "in the field."

As with any election night, it is a fluid situation - but when Fob James decided to declare victory 30 minutes before Clifford had lined up an exclusive interview with Candidate James - I met Cliff Hill for the first time.

He was rounding the corner at the Holidome in Montgomery - Raker in tow - and both were hollering “we’re f%$^ed!!! He’s declaring early!!!”

I was 21 years old, and thought it was the funniest thing ever.

They disappeared into the production truck that Donny Seymour had driven down - which was patched into a Crawford uplink truck from Atlanta. They emerged moments later and still got their interview (which we stole in Tuscaloosa thanks to the satellite) and all was well. But I learned then that “Cliffy and Jeffy” were tough customers - and I was glad to work with both of them full time less than two years later.

==============

Dick Clark’s bloopers and practical jokes - seeing Gary Dobbs on the key wall with that fantastic shot of the bridge in the background. He’s rattling off current conditions, and then the bridge is destroyed in a demolition blast! His reaction was priceless! Hope to see that clip again ...along with “picture perfect panoply prognostications.”

Debi Benson Bradford – Programming & Christmas Parade Organizer

A WAAY-TV Christmas Parade Memory & Updated Programming stories.

What a memory…*[Editor: - you can't make up stuff like this.]*

The WAAY-TV Christmas Parade was a signature event for the station. Everyone got involved, and tens of thousands of people participated each year. (I guess it still is and I guess they still do, but I digress.) About 10 horse groups usually participate. We placed horse groups at the back of the parade for obvious reasons. This particular year was going fairly well until we saw an uninvited "gentleman" riding a longhorn steer somewhere around the front half of the parade. Obviously drunk, the man was weaving and waving his cowboy hat to the astonished crowds while riding bareback on this giant steer. Throwing his head from side to side, the huge animal was snorting and looking like I would imagine any disgruntled steer would look. (Did I mention it was a longhorn?)

Mothers were yanking children off the curbs, and everyone along the route was warily watching, not sure if this was planned or not. Our inebriated urban cowboy was "guiding" this HUGE animal (a longhorn…oh, did I say that already?) by a single rope attached to a ring in the poor steer's nose. He'd yank the animal's head and blood would spray. We were horrified! My trusty horse committee and our HPD policemen friends tracked him down and removed him from the parade after a couple of blocks.

The rest of the parade went by smooth as silk, and I thought the whole thing went well except for the longhorn incident. However, Monday morning I received an onslaught of calls from irate parade attendees about yet another incident involving a very large animal.

Toward the end of the parade route, at the spot where the horse groups disbanded (and parade goers parked their cars) people had watched in horror as several drunken horsemen threw a horse down on the ground and castrated him! Again, blood was flying with the animal, and children who were watching, screaming. One family volunteered to come to the station, watched the raw tape, and identified the offender. Turns out he was a fire fighter from across the street! Turns out, the longhorn rider and horse castrator were buddies from the same (drunken) horse group. The group was banned for a year and the entire incident was certainly a "memory maker."

UPDATE 6-20-03 - PROGRAMMING STORIES

Programming memories:

My first job at WAAY-TV was Assistant Program Director to Bettie Higgins. My desk was in the back of the front lobby, and behind me was a large display of large, matted TV Guide covers. TV Guide sent these prints to the affiliates whose networks programs were highlighted on the cover. At that time, "Greatest American Hero" starring William Katt was one of the number one programs on network television and the TV Guide cover with Mr. Katt in his hero costume was prominently displayed behind me. During the summer months, neighborhood kids would drop in to buy a soda from the machines or get water. My first summer, three little boys propped their bicycles outside and came in for their usual rounds. One of them, the obvious leader, was showing the other two boys around the lobby before taking them down the hall. They took stock of the TV Guide prints and talked about their favorite shows – especially "Greatest American Hero." The leader of the pack puffed out his chest and gestured with his thumb toward the studio door. "Yep," he announced with authority, "they film 'Greatest American Hero' right in that room!" His friends, incredibly impressed, were then led down the hall by our little expert to get their soda.

People were rabid about ABC Nightline with Ted Koppel. We aired Nightline at 11:00 p.m., sometimes 11:30 p.m., depending upon programming. One day I received a call from an irate lady. Nightline's program the night before had some sort of sexual content or nudity buried within its content. This lady ranted and raved about "how dare you" and "who do you think you are" and "you're contaminating the minds of young children," etc., etc., etc. She went on and on and I could barely get a word in. Finally, I told her that the program began at (at this time) 11:30 p.m., and that particular section of the show hit around midnight. Surely, I said, children aren't up at that time of night and if they are then there are more important issues at hand – like why they're up so late. Well…she didn't like that, and said she was SO angry because HER son was up watching and HE was very upset which upset HER! Hmmmm…So I asked, gently, why she allowed her son to stay up so late to watch television, and she yelled back at me "He's 21 years old and can stay up that late if he wants to!!!" She finally hung up on me when I couldn't stop laughing.

Once I found a card that said, "Television is like your great uncle Bill. Nobody likes him, but everyone watches when he drops his pants and dances." I bought that card and kept it on my desk. That card helped me tremendously when rude and obnoxious callers began getting to me. Especially with the people who would tell me every little detail about a program that they didn't see fit for air, but they had obviously been watching for a long time. I quoted that card to a particularly obnoxious person once, thinking a little levity would calm him. He hung up on me, too.

Shawn Jarrett – Dan Quayle in Town & Other Jokes '88

Lew Koch hired me in March of '88 as an AM studio camera op, despite me showing up for my interview wearing camouflage pants and a teal-green button-down. (It was laundry day.) That was a pretty bad outfit, even for the '80's. Anyway, I'm still here at WAAY; in other words, the chunks of ice falling from the tower haven't gotten me yet.

The first really memorable assignment I had didn't involve shooting commercials (which I still do) but serving as a grip for Eric Eisgrau when Vice President Dan Quayle came to town in 1989. We all used 3/4 gear at the time, so I carried the deck, which was connected by an umbilical cable to Eric's Ikegami tube camera. (I really don't miss those.) Later that year, I would do some of my first solo shooting the day after the Airport Road tornado in November. It's hard to remember a bigger team effort; everybody pitched in to blacken the eyes of the other two stations, who learned the hard way that big diesel generators can be handy to have around.

One thing this building's seen plenty of is practical jokes. From paging someone to "Call 89" (which activates the intercom), to slipping bogus names in the Waay Too Early contest bin (ask Gary Dobbs about our friend "Jenna Taylia"), to finding a live sheep in my office, I've been accomplice to/victim of some good ones. In fact, I almost became a casualty early one morning in 1990 as the result of a joke; Anne Yoel did weather graphics for the AM news, and came in a little earlier than me. I walked into the dark studio and saw Anne rushing at me with a can of pepper spray. She saw it was me (thank God) and told me that "a man was hiding in the studio and yelling" at her. She was absolutely terrified.

It turned out that Tim Clemons, then a director, had recorded and looped a cart in which he just said "HEY" every 15 seconds, and ran it thru a speaker in the studio. Well, it worked, and the fact that I almost paid the price didn't stop me from ribbing Anne about her "stalker" for a long time.

Jan Miller – Operations Coordinator, What Station?

One of my earliest impressions of life at WAAY-TV was formed when I was newly hired in 1979 as a part-time typist in the Traffic department. My job was to type the log as fast and accurately as possible. (We didn't use computers back then as Mark Derrick was still getting the Radio Shack TRS-80 computers up & running.)

Sometimes I would finish early and had extra time, so I was asked if I would like to answer the telephone at the receptionist's desk. Of course, I was happy to oblige, but I was a bit nervous as I never knew what questions the callers would ask or how I would answer them.

One day, I was sitting at the desk in the lobby just outside the Traffic department when the phone rang. I answered it and said, "Good afternoon, CHANNEL 19" at the exact moment that Vice President & Director of Sales, Dan Whitsett, walked by. Without hesitating he said, "Oh, I thought this was CHANNEL 31," and walked on. I was mortified and was sure I was going to be fired immediately. I kept waiting for the bad news which never came and from that time on I knew I was really going to like working at Channel 31 because of the terrific people there, like Dan and all the rest.

There was such a variety of personalities yet the camaraderie we shared made working at the station so enjoyable.

Jan Miller (1979-1986) (one of Lew's Operations Coordinators)

Bill McClendon – Announcer & Audio, Fired and Re-hired

[Note: This is M.D. Smith telling of a story that Bill McClendon recalled during a visit to my office on 6-6-03.]

My father, Mr. Smith, III, as we all called him, was prone to occasionally mix one of his favorite pastimes, talking on Ham Radio with another, a bit of Canadian Club. After an almost all-night session, he would sometimes come up to the TV station after just an hour's sleep, but shaved, fresh white shirt and nice blue suit. However, he was also likely to issue orders to some staff members directly, without my knowledge (V.P. and General Manager). He even fired some people on occasions, but I would have to reassure them to wait until tomorrow to take it seriously, that I was sure he would reconsider (if he even remembered). Sometimes I could hardly tell that he was influenced by "leftovers" on days like that. The regular staff who didn't know him at all, certainly could not tell. But as Bill McClendon recalled, there was a sure sign.

When he begin talking to a person (Bill for instance) he'd begin with a bit of a "click" you make out of the side of your mouth and then say, "SPORT . . . ." and whatever. Well, whatever else followed you had to learn to just disregard. As Bill recalled, he was relieved when he found out he was NOT in trouble after the word "(click) SPORT" was said, but Bill also said he lost a few raises that way, too.

I hope Bill will follow this with some other classic stories tucked away in his memories from the Radio and TV days and his many dealings with the "House."

Jim Travis' Story & Pictures, Coffee Break, Primitive in '69

The photo is from 1970. Jim was News Director and anchor for WAAY-TV in 1970. Moved to Nashville December 1970, worked ten years with the ABC affiliate followed by 21 years with WSMV-Channel 4 (NBC).

Jim Travis with Bill Vance, then chairman of the Alabama Democratic Party. Vance later became a federal judge. He was assassinated at his Birmingham home by a letter bomb.

From TV Guide, May 16-22, 1970

May 16, 1970, TV Guide full-page ad for Coffee Break with Johnny Evans and Mary Beaton. The 30-minute program appeared Monday through Friday at 9:00 a.m. Cover article in this issue was by Vice President Spiro Agnew, advancing his criticisms of network television. 12 Feb 2003 Travis on CNBC's Capitol Report re: Sen. Fred Thompson decision not to seek re-election.

(More submissions from Jim)

Jim Travis, WAAY-TV News 1969-70.

My first impression of Channel 31 was not very good. It was the week after Christmas 1960, long before the Smith family took over. Huntsville's population was still under 100,000. Our family had just moved from Oklahoma. I had a little trouble following the thicker accents. In our unschooled minds, Bud Wilkinson was the great man of football. We didn't know much about Paul "Bear" Bryant and couldn't figure why 'Bama had an elephant for a mascot. That would change. In short order we learned to forget the elephant and just count the victories and have pride and faith in the country's new premier football team.

On Channel 31, a local entertainer did a good imitation of a mule. The late newscast went 15 minutes. The anchor also had to do a live commercial. When he finished, the screen faded to black and he would reappear in the anchor chair. We wondered how he did that so fast.

One night the screen wasn't blank long enough. We caught the newscaster jumping feet-first into his chair to finish the news. Now we knew the secret.

That was Huntsville TV in 1961. We had a good laugh and bought a big antenna to watch Nashville and Birmingham stations. They were more like what we'd been used to in Tulsa.

Cotton balls dotted the roadsides near our house. Rickety green city buses didn't run very often. Windows rattled from rocket tests miles across town. That was the extent of Huntsville excitement. I soon left for college and got into radio, never imagining that eight years later I'd drive up Monte Sano Mountain to start TV life at that station.

It was in Anniston, around 1963, that I first met Adrian Gibson. Even then his can-do attitude was an example for the rest of us.

My first week at what was now WAAY-TV was in June of 1969. I was led to a studio set to anchor the early news as fill-in. It was the first time in front of a camera, except maybe for an audition.

One of the studio crew showed how to string the microphone. The mike was the size of a salt shaker, which is what it was nicknamed. The lights were blinding and hot.

By the end of the first newscast, a wooden performance, I doubted the job would last a week. I would surely regret leaving radio. M.D. must not have watched too closely because he kept me on. The TV life stretched another 33 years.

You had to work in newsrooms of the 1960s to appreciate how primitive they were and how today's equipment is mind-boggling by comparison.

Five people made up the news department: two reporter/anchors and three photographers.

M.D. IV usually did the weather. His maps, covered with symbols, would arrive on a fax machine wired to a shortwave receiver. *[Editor's note: The Fax maps came on a machine wired to a telephone line and we had to pay for that, but the hourly sequence reports, temps, forecasts, etc. all came via teletype hooked to a shortwave receiver tuned to international Weather Service. It was free and therefore good.]*

One night I was anchoring the 10 p.m. news, M.D. didn't show. Frantic, I ran to the newsroom during a break, grabbed some weather copy and read it, half out of breath, standing off-camera away from the green map. It was a cold, wet night. As I mentioned rain, a steady drip fell from the ceiling, in front of the map and to the floor. Anyone at home could have seen it. It was the first and last time I ever did TV weather—the last time I ever wanted to.

Dan Akens had left WAAY-TV for NASA some time before. I met him a few times and was introduced to his witty caption-writing. I might have caught him doing news, when I worked radio in Anniston as a college student. As fate had it, in the next twenty years we would work at the same TV stations twice. I never seemed to shake him for long. Dan's time at NASA and later at WSB-TV evidently turned him into one of the best "big story" coordinators I ever met, terrific at marshalling resources.

In the late sixties, TV newsrooms were switching to color film. Channel 19 had made the switch by 1970. We still used Tri-X black and white reversal film, developed in a small Kramer processor in a closet near the big studio. All of us learned to use it but Eric, then chief photographer, did much of the processing.

The sole film camera was a bulky Auricon. Everything was heavy—the camera, its battery, the lights, their batteries, and the tripod.

Light bulbs would often blow just as an interview or a news conference was beginning. That could be embarrassing if you were at Werner von Braun's news conference with the world's press lining both sides of his long conference table.

Film costs were budget-killers. After becoming news director in 1970, I recall a meeting at MD III's home in which we all discussed the possibility of going straight to video tape. It would be expensive but would cut out film entirely. The downside was that existing portable video gear was even bulkier than film cameras. It was used mostly on network sports events, in black and white. As with other stations, WAAY would not switch to video cams until years later. What impressed me in 1970 and still does today is how far-sighted the Smiths were on this.

Two advantages we did have were the twin-engine plane and the boat, which M.D. was quite willing to use for a good news story.

Through the years, a fairly large body of tales has developed around the station and its people. For me it was TV baptism. It was a great time to live in Huntsville, when folks at Marshall helped put a man on the moon and would celebrate some of their successes with rallies at the courthouse square, events we covered live with equipment borrowed at least once from a church remote. Whatever was needed, Cactus would find a way.

That period is as memorable as high school is for most. Our first TV or radio jobs, our first paying jobs, folks who give us the chance to make good, are what we remember best, next to memories of our families, because they become as family.

We all start somewhere. WAAY-TV was as good as any, the experience as valuable as we worked to make it. I remain grateful for the year and a half I spent there. No less satisfying was learning in later years how WAAY-TV News had grown in size and audience.

There have been lean times and good times. The real job, to give viewers good information, has not changed one bit. One thing for certain: we never want to go back to those days 40 years ago when a station could get by with one news reader, who also did a live commercial, who occasionally was caught jumping back in the anchor chair after a fade to black.

Jim Travis - Dickson, TN June 21, 2003

Cliff Hill – Anchor, Reporter Now Lawyer – Getting Hired

To my WAAY family,

It is Sunday morning at the law office (6/22/03) and I'm writing this after driving down memory line with stop and go traffic through the many stories on the web site. I now want to add a word or two about my dream to work with all of you.

I started watching Channel 31 a long time ago, I mean a long time ago. We would visit my grand-parents during the summer years for a couple weeks as my father was station with the Air Force. We would all gather and watch the news and other programming after the family communal dinner. It was a blessed time with generations of Hills solving the world's problems. I remember thinking this TV stuff was cool and it only took a couple of "eyes in front of the set" visits with Johnny Evans on Saturday mornings to hook me. My sister even appeared on his show. She was the star of the family for years.

Fast forward, to the beginning of my broadcast career. I set out with one purpose in mind and that was somehow, someway to get a job at WAAY. No kidding. Those who worked close with me in the field know what I'm talking about. I sent tape after tape to the station for years. Finally, I got a call from Cliff Windham. Cliff called me and we set up an interview date. I came up and did the walk around thing with Cliff and met with Mark McGee. Finally, I met the man. M.D. visited with me for a couple of minutes discussing the usual stuff and then he popped the question. "Cliff, how would you do a better job reporting on stories than the competition?" (or something similar to this in any event). Well, I remember it knocked me to the floor. I stammered saying something not really bright like, "Well, Mr. Smith I'll sure work hard." Of course, I didn't get the job and was down but not out.

I set out on a new course and for the next couple of years I decided every time I got a vacation or was in the area I would call and drop off a tape. I wouldn't take the hint after that didn't work. So, I started out on a new strategy. I figured I could get a moment with Cliff Windham by calling him and saying, "Hi, Cliff this is Cliff." I would usually get a couple of minutes with Cliff. Then I would get to drop a tape off to Mark. (Years later Heather Burns found one of those early tapes of mine and let's say, I know why I wasn't ready for WAAY.)

I finally made it to Huntsville when 48 brought me down from Cape Girardeau to set up its morning show. (The show was cancelled 9 months later.) Then out of the blue came a call from Mark and Cliff who needed a fill-in anchor for the weekend. That call ended my long quest to work at WAAY.

Through the years at WAAY, I had the honor to work for and with many fine people. Channel 31 was a family. As a family we sometimes got into arguments, we shared each other's ups and downs and through it all did some really good work. I enjoyed the support I received from my co-workers and the management relative my investigative work. The man in the corner office, while reserved and cautious about my investigative stories would throw his full support behind them when the decision was made, they would air. And even more importantly he stood with me when we ruffled more than a few feathers.

Years later I asked M.D. about his question to me as a young aspiring reporter. He saw so many come through the doors it didn't surprise me he didn't remember the question. There really isn't a right answer to the question. It was a question reflecting one's ability to think on their feet, a necessary requirement for live television. Before the doors closed on my time at WAAY, I answered MD's question, saying, "Simply do the story better than the competition no matter what it took." I believe that was the standard at WAAY during my time. When I lived up to it, it was the best of times, when I didn't, it was the worst of times.

Cliff Hill

[Note: this message was sent by Cliff AFTER the 31 Alumni Reunion Party.]

MD,

I know you're not going to see this until the party is over at whatever time tonight or tomorrow morning. However, I want to thank you for putting this together. I've called across the country looking for my friends from by-gone days and alerted them. You have a great big family at home and at the channel 31 of yesterday. I was proud to work for you, honored you hired me, and grateful for your support along the way. When the sale of the station was announced and the handover took place, I was quoted in the paper saying, "I was proud to be going out with you." It was true then and true today. You've meant a lot to all of us and continue to do so.

Thanks for the party. Should you ever get another station call and I'm there.

Cliff Hill

Rob Weske 1997-1999 (Promotion) Funny Guy & Stories

A New York Yankee in WAAY TV's Court

The station on the mountain provided me with a unique experience in my career in television. I've worked in markets ranging from a tiny CBS affiliate in Binghamton NY to KNBC in Los Angeles, California. So, I've been able to work from sea to shining sea, and my time in northern Alabama was chock full of southern hospitality. As I am a former resident of New York City, I thought I was prepared for life in Dixie. The proactive word here is "thought."

-- My first week at WAAY, as I was converting my New York State license and plates to Alabama's, a thick accented woman at the DMV told me, when finding out that I worked at WAAY, I needed to get "Ruy-sin In The Sun." I had trouble understanding what that meant: is that a local beverage? Suntan oil? "No, the feeelm! You need to show the feelm!" she replied. The classic movie—"Raisin In The Sun,"...she wanted WAAY to air the movie again. "Hadn't seen it in years! Y'all got to air it again," she assured me. I nodded as I grabbed my Alabama license.

-- Weeks later Yodi Lockhart alerted everyone near my desk that Bill Barry was "handling snakes" outside Lew Koch's window. True enough, as we stood in Lew's office and watched with awe, outside the window, on the side lawn, was Bill Barry carrying one of the biggest freaking snakes I've ever seen. Welcome to Alabama. I was assured it was only a "rat snake," and they are good for gardens. Yeah, right. So are lions. And I quickly heard that there's some of those up on the mountain, too. Nice.

-- I was indoctrinated to the fine cuisine at the Double Springs Cafe--(The "double" means "double wide"—it's a trailer that doubles as a diner.) Anyway, Tracy Slayton and Rob Boenau brought the "new Yankee" into the Double Spring Cafe for lunch. I was prompted on what to order...the burgers. Burgers, I was told by the waitress, are served in 2 ways—with "grease" and "double grease." Well, my two southern hosts could only wince with embarrassment as I questioned the waitress about what "grease" and "double grease" really meant. I committed further stupid blasphemy when I dared ask if they had anything other than "RC Cola."

-- Shawn Jarrett, the Prank-master General, welcomed me early as we drove off to a promo shoot. He made sure I was seated into the "Chair Of Pain"—a precision instrument of shock and awe, used at WAAY before the military picked up on that term. Shawn is a top caliber prankster. I've seen him cover an unaware coworker from head to toe with a sticker-tape parade of paper. I've seen Shawn customize license plates...stash plates of Gibson's BBQ in nooks and crannies of offices that you'd never think were there...I've seen him even doctor a coworker's personalized Star Trek episode in ways that would make Industrial Light & Magic proud. The master prankster is so skilled at conceiving and sniffing out a prank—my one foray into pulling off a classic prank on him had mixed results due to his instinctive powers; don't ask how, but a live sheep was obtained and transported to the station in the early morning hours. We placed the sheep in Shawn's office, and right across the hall, several of us were out of sight, waiting his arrival with a rolling camera, ready to ambush. As Shawn approached his office, he was incredibly wary—he always had a sixth sense if a prank was bearing down on him—he'd feel the joke before it reached out and bit him. As he put his key into his door, the sheep made a rustling sound, and Shawn coolly stopped what he was doing, and continued down the hall, away from the office, like nothing was out of the ordinary. There would be no camera ready "ambush"...it would STILL have to be on his terms. At least the master, when introduced to his sheep, appreciated the imaginative effort.

-- Chasing storms: that's something you don't do in the course of the day in other markets—but at WAAY, it is part of the job description for reporters, photogs, and promotion people. Some incredibly powerful storms tear through that part of the country, and our shooters would think of nothing to put themselves in harm's way. The photogs at WAAY had more guts than photogs I've worked with in any other market, even market number 2.

-- Ben Boles' Pepito's pilgrimages. That guy was never shortchanged at that place.

-- The little Red Devil herself, Toni Harper. The Sunflower queen always brought mischievous sunshine into our days, especially during football season. Everyone needs a Harper in their workplace.

-- Promotion Department co-workers Bob and Inga...we had a lot of laughs. I couldn't have imagined my time there without them.

-- Eric Eisgrau—known as "Mr. Sunshine"—and also known as "Mr. Dependable" when it came to video for our deadlines. He always made a Yankee feel at home…even though he'd use the word "damn" in front of it when addressing me.

-- And many more who left an impression on me up on the mountain. There aren't too many TV stations in the country that had the local flavor and warmth of WAAY TV. It's all about the people you're with. They make it WAAY worthwhile.

Kim Albright – Hard News Hound & Dick Van Valkenburg

The first TV person I ever met was Dick VanValkenburg. I was 19 years old at the time and just starting my broadcast career at WCNA radio in Scottsboro. Cactus was the engineer for the radio station and had given me Dick's phone numbers in case I ever got anything "good" in Jackson County. We had a bazaar bank robbery in Stevenson and I called Dick. Fast as lightning he drove from Huntsville to Stevenson. I met him on the road and guided him downtown to the scene. He and I spent several hours together that day. I don't know what he ever saw in me but for whatever reason Dick seemed to take a "likin'" to me. Over the course of the next 14 years or so Dick and I forged a strong friendship and he taught me most of what I know about spot news. He was truly the newsman's newsman, the one everyone strived to be nearly as good as, although nobody ever came close. After Dick's health declined and he couldn't run like he use to he would call me to be his eyes and go "check out" potential stories.

After Dick passed away in December 1986, I received a call from Cactus that M.D. Smith wanted to talk with me. I cannot describe the mixed emotions I felt. I was very sad at losing my good friend Dick but I was being given the chance to live my dream to be "the" spot news guy for the #1 TV station in north Alabama. Over the next 12 or 13 years I met and worked with some of the nicest people in the industry...Waay too many to name. I know at times during those 12 or 13 years I probably got on a lot of people's nerves including MD's and we all had out disagreements, after all we are all humans with our own opinions, but MD, Mark McGee and Cliff Windom were always there willing to listen to my ideas and take time to explain why they wouldn't work if they didn't think they would.

Having been away from "the family" now for almost 4 years I can truly say those 12 or 13 years were the very best years of my working life. Yes, the hours were long, yes it took me away from Lynn much too often but looking back I'd do it again tomorrow if given the chance. I learned so much...how to get microwave shots from "impossible" places, how to do live traffic reports, underwater reports, live in aircraft reports. We tried and accomplished things as a team other stations are not willing to even attempt, even today! That's what made us #1 and what made us a family!

Kim C. Albright

M.D.'s Birthday Surprises to Remember – Margaret Forrest & Priscilla Cowing

The staff always loved creative pranks of all kinds. That's one of the things I have to say I miss most not being in the TV business anymore. The creative juices were always flowing from a lot of the staff, particularly the production and news department.

I don't know whose idea it was for Margaret Forrest, who did singing telegrams at the same time she worked in production, to greet me that day.

I was ushered into the studio on some ruse of needing to see a leak in the building. (So what else was new?) When I got thru the studio door, I see most of the staff gathered there, and knowing it was my birthday, I blushed knowing something had been planned and was going to "surprise me."

As the staff began to sing "Happy Birthday" to me, Margaret (dressed in her French Maid outfit) came through the hall door into the studio and presented me with a gift and later a cake. I forget what was in the sack, but it was a gag gift of some type for sure.

You will see the 1988 set of "Morning WAAY" behind us all decked out for the Christmas Season.

Another memorable Birthday "trap" was held in the new conference room where I was summoned by Melody Rice (so it was early 90's) to come and I was suspicious and said I was not coming with her. She begged me and said even if I was right about another "surprise," I could not disappoint the people. Well, sure enough, as I walked through the door, there were all department heads and a good part of the staff standing behind the table and a lighted birthday cake was on the end of the table waiting for me. By then, the candles were just about burned down, as Melody had a hard time coaxing me to come there.

But then, the crowning moment was the voice of Priscilla Cowing belting out "Happy Birthday, Mr. President" in her best Marilyn Monroe voice. She did quite an acting job and everyone was greatly amused, except Mr. President who had a red face. (Seriously, I later always enjoyed the fact that the staff cared enough to go to such trouble on my birthdays.)

Tim Hall - 1992 Panoply Story & Trouble

I think it was Panoply 1992. When we did something at WAAY-TV, we did it RIGHT! And Panoply was no exception. We set up a stage, lighting, the works! Boy did we look good. It was one of those Sunday 5:30 newscasts I was anchoring with Amy Witte. During the first break...something happened. The lights went out. I jumped up out of the chair (in my highly unusual manner to get excited about anything) and began yelling..."We're dead in the water. We have no power. We're dead in the water." Guess what? We were on the air the whole time. The only thing that had happened was someone had tripped over the power cable for the lights. So, there I was...flapping my arms like a bird yelling in front of everyone.

The next week I went to church. Two little girls walked up to me and said, "Hey Mr. Hall, We're dead in the water. We have NO power!" I looked at them and said, "Shut up!" Their mom got mad at me and said, "that's no way for a preacher to talk!" The stupid mistake got me in trouble two times!! At work and church!!!

Working at WAAY-TV was a highlight of my career in TV news. You made it fun and a great learning experience. I miss the days of the family environment and hope everyone will come out to the reunion. I look forward to seeing you all!!!

Ken Rainey - A Deliverance Trip with Cactus – Wild

My first official day on the air was July 15, 1974, but I arrived in town early for the usual reasons, to get temporary lodging and to check out the housing market. My first weekend was coming up, when MD ask me to come along on an all-boys outing on Saturday morning. He picked me up early and we drove east to hook up with Cactus.

Cactus owned some land on top of a mountain somewhere around Scottsboro. He also owned a 4X4 Jeep with a winch, (a winch is not a woman, but a mechanical device that rednecks put on front of their redneck vehicles). As you might guess, this motley crew was armed to the teeth with 22s, bourbon whiskey, and beer. Food you ask-forget it. I must admit that Cactus had a nice place, heavily wooded, several caves, and the remains of a very old log house. (The walls were still standing.)

First order of the day was a tour of the property. There were no real roads and the ride was so rough that my beer kept foaming up. That was followed by target practice, more drinks and a continuous rendering of the dueling banjos from an eight track. ("Deliverance" had recently been released in the movie theatres.)

When the shooting stopped the conversation drifted towards making money. Cactus was convinced that there was a big market for old timbers that could be used as fireplace mantels and that they should be worth at least $150 each. So, there we were, three fools using the winch to pull down the walls of the old log house. Cactus said he would collect the timbers later because darkness was upon us and it was a long, dry, drive back to civilization. Dry? How silly of me. Cactus knew a local bootlegger and we were able to pick up another pint.

I'm not sure how that day ended except we know that MD went home to Judy. Cactus likely went home to pass out, and me, well it was Saturday night and I didn't want to miss the action at the Hour Glass.

I don't think the three of us ever mentioned that day again. Till now.

[Editor's note: Ken's exactly right about that story. Cactus and I agree on it happening just that way. It was a great trip. The timbers Ken is talking about did get used as Cactus said they would. Below is a photo from Cactus' den at home, which he added in the fall of 1974, a few months after the trip in July that Ken remembers. Cactus says it is "wormy chestnut" and made a great mantel that is in his den today. He stored a lot of the other wood and eventually it has ended up in one of his daughter's houses in the country outside of Scottsboro and other places around the area.]

April 3, 1974 Tornadoes & Damage – Mr. Jordan's Book

[Note: all quotes and photos from Mr. Jordan's book are used with specific permission by the author in Moulton, Alabama. 6-26-03 Any other use of this copywritten material is forbidden.]

These are some passages from the book, "April 3, 1974 - A Night To Remember" by Charles Jordan. It's the most complete printed works of those tornadoes in North Alabama ever printed. He has many photos, like those of Gov. George Wallace in a wheel chair visiting the devastated areas the next day.

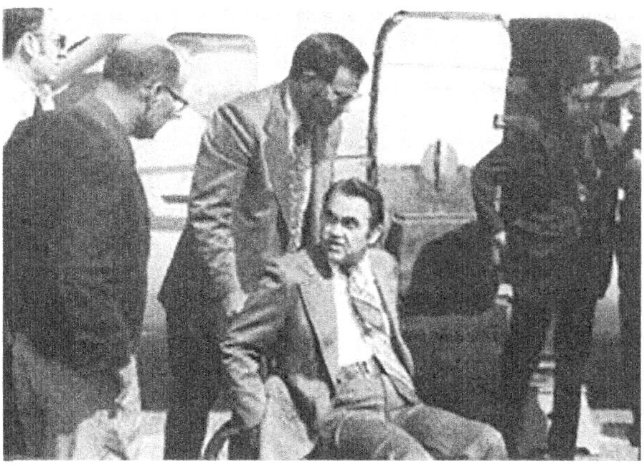

Tons of photos of the destruction. There are views from the ground and air. Personal accounts of about a hundred eye witnesses from all over the state. We were the only TV station mentioned in his book, primarily because we had the only weather radar in the area. Here are some quotes:

[Note: at least 6 tornadoes moved across Madison County in the course of that day.]

"WAAY-TV, Channel 31, zeroed in on its radio scope and the twister was cutting a savage path into the county. There were those who were becoming convinced that the television station was over-dramatizing and that the situation could not possibly be as bad as reported."

"Again, Huntsville Manufacturing sirens and loudspeakers blasted out warnings. An eerie stillness prevailed outside. How, some wondered, could another be on its way when the clouds appeared scattered, the winds calm?"

"It was 10:50 p.m. in Huntsville when soldiers at the Army Missile and Munitions Center and School looked out through the blackness to the saddle of Wheeden and Madkin Mountains to the west."

"A mammoth twister was working its way through the saddle, bearing down on the troop area where more than 1,000 U.S. and foreign troops were housed for missile training. There was little more time than to move to the bottom floors of the multi-story barracks and drop to the floor…More than 98 building were either destroyed or severely damaged, but no one was killed…The twister struck again near the Patton Road and Drake housing area…MacDonnell Elementary School was demolished…It smashed into Glenn'll Trailer Court in Southwest Huntsville…then on a northeast path on Leeman Ferry Road where it cut down everything in its path…At Sterling Cadillac (Drake and Parkway), the twister gutted the building…Flipped cars on Memorial Parkway...bent utility poles to the ground…slashed into the International House of Pancakes on Drake and other side of Memorial Parkway…then Parkway City where it ripped much of the southern part of the shopping center apart…Then into Thorton Acres where it ripped roofs, topped trees and smashed homes before moving on."

[Note from MDS: it seemed to lift up and passed right over Covemont where MDS III lived on DeRussey and MDS IV lived on Fairmont . . . the tops of huge pine trees were snapped off, but not much damage to homes.]

"Then on to the base of Monte Sano Mountain. It tore up the forest on the city side of Monte Sano, passed over Monte Sano Blvd. where Panorama crosses at the southern part of the mountain. That's when all the TV stations lost power at 11:05. Sarah Batson's little A-frame house was pretty much gone, but Sarah was hardly injured…Sgt. Owens brick house was rotated on his foundation, the roof missing and sat crooked the next morning. The apartments across the boulevard were badly damaged as were other houses in that area. Oak trees across the road blocking emergency crews. The tornado continued for another 18 miles, demolishing woodlands all the way to Madison County fishing lake."

Here is the list of counties where people died

Madison county 16
Limestone 16
Morgan 0
Lawrence 16
Fayette 2
Cullman 1
Winston 5

Hundreds more were injured and many hospitalized.

Another interesting fact...Huntsville had a small tornado the night of April 1, just before the big ones. It tore down the WAAY AM center tower of the 3-tower array and the disk jockey called Cactus at home and told him. Cactus' first reaction was, "Oh Sure, April Fools, right?" It took a while to convince him it was true. Then Cactus called me and I had the same reaction. By the end of the day, I was tired of my kids pranks and jokes on April Fools. I kept thinking Cactus was fooling me, too. We had to rebuild the tower and the tower that came after that, was very strange looking indeed, with the base of the larger tower, but a smaller self-supporting one rose back to the 125' level of the other ones.

So now, tell me YOUR story. Many of you were involved with the 1974 twisters as well as the 1989 one that tore up Airport Road apartments, shopping centers and finally Jones Valley School and that area. Pat Spurrier stopped by today and was telling me of the phone calls she was answering the evening in November 1989 from people wanting to help. A man from Georgia called. She also recollects the Ice Storm of February 1985 and the days without power trying to manually do logs and be ready for when power came back. (We installed our generator around 1987 - two years after my father died - he would have really smiled to see that because of that forethought, we were the ONLY station on the air that November in 1989 after about 5:30 in the afternoon and we had 100% of the audience. It was so high; the other stations made the rating services take OUT our ratings that evening to get numbers they could live with.)

Photos from Mr. Jordan's illustrated book: Gov. George Wallace, Sterling Cadillac building gutted.

Sterling Cadillac shows fury of storm

Sam DePino – Anchor & Newsman 1972-1974 – His Tales

[Note: Sam's original comments were 7500 words long and too much for this book. Only selected comments will follow with # # # between sections that are included.]

Happy reunion, everybody. Good idea, MD. I see you're still doing the same thing, getting people together. It was that skill, your family's commitment to becoming the best news operation in the market, great advice from Phil McHugh and an on-air team from 1972 to 1974 with the chemistry together that audiences were looking for. You had outstanding people behind the scenes and we all won. It "looked easy" but it wasn't. The challenge of keeping that audience became the task of a new consultant and whoever stayed or left or were yet to come. I moved back to major markets in broadcast news and in New England and Washington, DC in media related fields.

My greatest memory and greatest source of professional satisfaction came in Huntsville. It was during the April 3, 1974 tornado coverage. Sgt. Bob Owens told me we saved the lives of his family with our continued coverage. Some 50 people died in several Northern Alabama counties, many more were injured and the property damage I saw on the ground and in flyovers was immense all the way into Tennessee. There's nothing that tops being part of efforts that save lives.

Respectful condolences to those whose loved ones have passed away. Best wishes to the rest of you.

Sam DePino (more below)

# # # #

A few thoughts to share about a couple of touchy subjects. There are many, but in this and future "contributions" to the alumni, I'll try to stay with only the most significant. One is race.

The college picture and a quick feature on Good Morning America recently reminded me of some progress we made in Huntsville toward racial equity. One famous alumni is Robin Roberts of ABC's GMA. Much too young to be a classmate of mine in 1953, Robin attended much later on a basketball scholarship. The coach's wife sent her pic to GMA in early July 2003 and they showed it. Part of our speech course in 1953 was debating all sides of integration, which was destined to happen soon. Just out of Annie Eastman High in Hammond, Louisiana, I worked to earn my first year's college expenses in furniture manufacture in Grand Rapids, Michigan and remember being told, "Sam, we have integration in this plant and you could be being supervised by a 'colored.' Any problem with that?" None then. None since. Ask Garry Armstrong, former ABC editor with 35 plus years after in Boston and Les Brownlee, Chicago's 50 plus year journalist, now professor at Columbia College. Robin's preparation at SLU shows brilliantly. She is among a long line of very accomplished "people of color."

I'm not sure if I was the first to propose to MD that we get not just one African American but two. I used my "bottom line" approach, "sixteen percent of the potential audience is black. That would go a long way to increase our audience and having two would really beat the competition." We hired John Barret Townsend. I couldn't resist John's quoting MLK three times, "I have a dream" (to be in television news.) Soon, we hired Linda Ammons. Both were excellent workers and responded very well to the help and advanced training we gave them. I think they were significant in our efforts to win the ratings. (Sam's previous station in Mobile had no minorities on the air) MD was much smarter. One consolation in Mobile was that Hank Aaron's sister was the newsroom secretary.

# # # #

[Note: we had hired the consultant firm of McHugh-Hoffman in 1971 and they were one reason M.D. IV hired Sam.] My attitude "problem" remains. I'm still a journalist. I defended the consultants in 1972 in the

official magazine of the Society of Professional Journalists during a time when they were under heavy flack. This, at considerable risk of a good reputation I had earned before and during my membership. I said the consultants help a lot. Phil McHugh had EARNED my respect and I felt someone should speak up. I wrote QUILL and they published it verbatim. I'll paraphrase. I said even Joseph Pulitzer said to tell the stories in a clear and understandable way, making it interesting while being fair and objective. McHugh never interfered with the news content. Consultants help in research (which good reporters do anyway) to get a handle on what the audience is concerned about and they are outstanding at many other aspects of presentation, promotion and strategy. They bring ideas from other markets and know who's available. MD mentioned "talent search" as part of his reason for changing to Magid although McHugh had helped guide WAAY TV to the number one spot.

MD told me recently, "I personally liked Phil McHugh a LOT. I hated to change to Magid, but I needed the kind of research he would provide and the depth of tapes they sent showing how other stations do certain features, talent search, and a variety of things Phil could not provide."

I disagreed. Phil McHugh was extremely valuable to our efforts as I'm sure Magid proved to be when they came onboard. I looked forward to learning more but left for Houston soon.

The consultants were just beginning to be really effective in the late 60's with Phil McHugh leading the way. He was followed soon by Magid, Al Primo, and several other consulting agencies. I didn't recall anything in that oath or our journalistic ethic that said, "get those meddling so and so's outta heeyuh." I wrote that the consultants bring research, ideas, production and presentation values, and strategies for winning.

Of course, no endeavor worth doing should ever be done without give and take and agreements to disagree at worst and, at best, work together and share ideas. I wouldn't want to work anywhere that everyone agreed with me or where I was expected to agree with everyone. It's from differences of opinion, approach and just plain ways of doing things that we get progress. Total agreement works real well in dictatorships and progress in such an environment has a very low lid.

Although I had built very strong broadcasting and journalism credentials from small to medium to major to network markets and back, I'm probably the best person to give advice on "how to be a real journalist and be highly mobile." I was told by the owner of a small-town newspaper when I was a printer's "devil" at age 8 and had just gotten my first byline, "you don't want to be a reporter in a small market, Sam. A reporter isn't worth his salt unless he gets ridden out on a rail within two years." Damned if he wasn't right.

(Title) The following is for adults only

One night, the late and very great Bob Sullivan (our engineer who threw the switches) came to me and said, "Sam, there's a great feature on the daily electronic closed-circuit feed from ABC." "Oh? What is it?" I knew this was coming from a guy who drove a great sports car which I wanted to buy and he wouldn't sell, so it must be good. Sullivan smiled, "Take my word for it, it's one of those features you refer to when you say you want something people will talk about."

Hey, if it's good enough for Sully, it's good enough for me. "Ok. I'll schedule it for the closing segment."

"It's not for children to see."

"Hmmm..." I thought maybe I should look at it, but I started thinking about a disclaimer of some sort. I trusted Bob because he also knew what gets audiences. I didn't have time to look at it, but I composed a teaser for before the commercial that went something like this: "If any children are up now at almost 10:30 p.m., send them to bed. We have something after the commercial that they shouldn't see." I still hadn't seen it. Bob had such a glint in his eye that I knew if he liked it so much a lot of people like him would, too. I ALWAYS pay attention to technicians and stage hands. They're more in touch with audiences than a lot of us "nabobs."

MD, place commercial popup here.

ANCHOR: "Again, the following is for adults only...." I'm still wondering what we're about to see. I forgot the lead in, which was provided by ABC, but it went something like this, "A woman in (I forget where) is earning her master's degree in Political Science (or some such) in an unusual job." We get more in this report from ABC's (Joe Blow or somebody, I forgot the name...not Peter Jennings, for sure.)

We got more, all right!

Bob had already started the roll-up of the tape and as almost every time it popped immediately on the Chroma behind me and soon filled the screen.

The lady was paying her way strip dancing!!!!!! The Playboy channel wouldn't have done a better job. The g-strings and pasties were flesh colored!!!! Bumpity-bump, whirl-whirl, dip-dip, swoop-swoop, flippity-flop, all to the raunchiest strip music and loud drumming you can imagine. The lady looked as naked as a jaybird!!!! Needless to say, my eyes were glued to the screen.

I got more verbal flak from one viewer than any time in my career before and since! That includes all that time I spent covering integration and antiwar protests! I think MD just smiled. I don't recall his saying anything about it. A very religious and irate young woman called me. She was the daughter of a former German aeronautical engineer turned Redstone and Nasa Scientist after World War Two. He had designed the first vertical takeoff aircraft. I think he was retired when he became a hobby shop owner. His daughter made me feel like I was forever and unflinchingly pretty much banned from the shop, which I had visited often, "how can you expect parents to get their children away from the television set?" You don't want to know what else she said, but it was all very prim and proper. I don't recall any cusswords.

I suspect that one feature was talked about an awful lot in our market for a long time. I also suspect our audience grew quite a bit after that.

In the meantime, Sully and I were still laughing about it long after it ran. I didn't tell anyone about that call, so you just got an exclusive. Happy now?

# # # #

My Mensa membership is irrelevant and the description of me as lacking tact (that's the correct spelling by the way) is about 10 per cent accurate. Intelligence comes in many forms at all levels and I can point to many geniuses with low IQ's. The battery of tests indicated I qualified to join Mensa and another so called group by being in the upper .004 per cent. So what? There are many people with lower IQ's who know hundreds of times more about their chosen areas of study or experience than I. There are people of lower than average intelligence who have contributed a lot more to humanity than I ever could. So, let's get that straight here and now. I can screw up as well or worse than anyone else, most of the time much more brilliantly and noticeably. I'm sorry MD brought it up. I'm not diminishing anyone with high or low intelligence. We all have our contributions to make. The challenge is to see and bring out that genius that is in every human being.

Yes, I was responsive but not always tactful. Who is? Dale Carnegie was never in a newsroom but he had a good handle on human nature. I told it the way I saw it, but there's no way for anyone else to know exactly how much I was holding back. I certainly didn't comment or criticize that often. I figured after the newscast was over, each person would work for improvement and nothing needed to be said, specially a lot of nit-picking. I was very satisfied with the team.

As producer, I supervised for the 10 p.m. show. That was the only newscast I was in charge of and, since we all did our work independently, it wasn't always obvious who was in charge. Together, we had a smooth news, weather and sports program that I would have put in any major market with pride at least half the time. **The early 70's news product was good enough to quintuple WAAY TV's news ratings which just about everybody thought would be impossible so quickly.** *[Editor's note, that's exactly what we did, going from 10,000 homes in November 1971 to 50,000 homes in November of 1975.]*

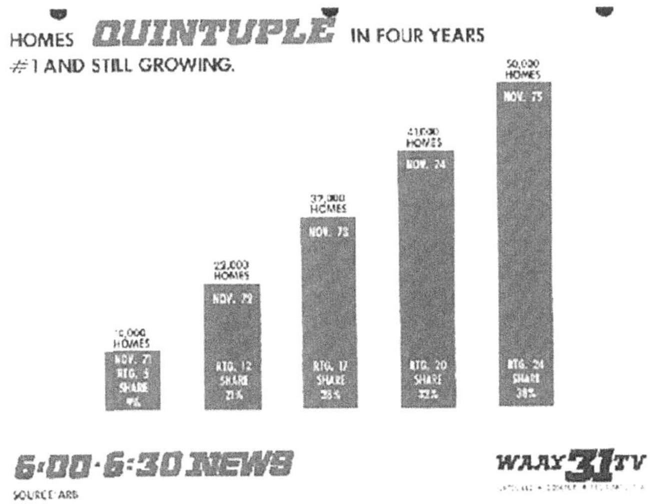

Technologically, we only had a more powerful radar than the competition with a sensitivity that at least one former meteorologist severely questioned. No Teleprompters. No live remote newsfeeds. We were rudimentary for the time. Such tools are fine and welcomed, but those of us working from 1972 to 1975 managed to bring Smith Broadcasting out of the cellar to the top without them. Such tools are just more resources to do what we're supposed to do, INFORM the public.

My "image" was what I was on the air. That was the true me. What they saw is what they got. That was the way I wanted to be with everybody. Many didn't give me a chance from the get-go. After all, I put my pants on one leg at a time, don't I? At worst, I'm very human. I would have appreciated that kind of respect. I always thought news people were supposed to be fair.

# # # #

A last word about "getting along."

If everyone agrees completely somewhere, something's profoundly wrong and there's a lot of intimidation going on. People are going to differ in their perceptions and approaches. It's those differences, when appreciated and taken in effect, which bring progress. Sure, experience brings us to the correct actions, solutions and fulfilling our common goals, but the thought processes in news gathering and presentation must have the advantages of input from everyone. This, no matter what their experiences or other preparation. Each person's opinion is valuable and should be considered if it's based on facts and not meant to harm. If each of us would compete with our OWN last performance we'd find continual improvement and perhaps as others see what we're doing, they'll try to reach the same plateau.

Answering all that was said about me as I appeared to some more than 30 years ago is an exercise in futility. People change. Whether a long past and shudda been forgotten impression can be changed is moot at this stage of the "game." I had put it all in the past until I noticed what was being written about me wasn't exactly what I would want to use as a letter of recommendation. I've said this, too: on balance, MD was one of several best supervisors I've worked with. He spent time with all of us, listened and selectively used everything we suggested or said, one way or the other. He was communicative, yet secretive as a good manager should be. He was a businessman. I was a journalist. I thought we got along "splendidly" too, considering journalism in the true sense is not a business. Business is guided by the profit motive. Journalism is motivated by public service without direct profit. I thought we had reached a mutual respect of each other's essential makeup. I was also a business man a few years, some before getting into news full time and some during my "vacations" from the profession after. I was also a salesman (on air and in a couple of job situations) before and after. After I left the Houston Television station, I worked in the Oil industry as a grunt iron worker and fabricator. I had manufacturing experience just after college and wanted to get back for a while as "bobby blue-collar" and recapture "his" qualities. I served as Northeast and Central regional marketing manager for a Silicon Valley firm and worked up to Vice President of Marketing within 3 years after I left channel 48 in 1979.

After all these years, I have nothing negative in thought or word to say about anyone at WAAY Television, regardless of what they may say about me. I accepted all of them as they were long ago. Change is something that comes from within. I've been green with envy myself. I understand.

James Hessler – Director & A Christmas Parade Hijacked

[Editor's note: a short story from James, about one of the Christmas Parades we televised.]

Christmas Parade Hijacked!

I don't remember what year it was, but sometime in the early 90's, the 31 Christmas Parade was hijacked, by Miss Merry Christmas of all people. We had set up two chairs for our hosts---Gary Dobbs and Priscilla Cowing---and partway through the morning, one of our weather graphics operators showed up dressed in a Miss Merry Christmas costume with some sort of puppet named was Poofy. I think the puppet was a sock.

Well, Miss Merry sat down in Priscilla's chair like she belonged there and we all accepted that she did. We just chalked it up to some small detail that somebody forgot to mention. Priscilla showed up and since we only had two chairs, we made Miss Merry stand behind Gary and Priscilla. Again, nobody questioned her right to be there. During the parade, she and the puppet interjected---or interrupted---freely. It just goes to show you what a little assertiveness training can get you---the role as a centerpiece in the parade broadcast! By the way, Miss Merry Christmas later became known, affectionately of course, as Poofy.

Kirk Chaisson – Investigative Reporter 1989-1994

I worked at Channel 31 from August of 1989 until May of 1994. When I left WAAY, I went back to Guam to become the news director for Guam Cable TV, which I did for three years, then I worked for seven months as a general assignment reporter for KLAS-TV in Las Vegas, before taking this job up here, to create and serve as the lead reporter for a statewide newsmagazine program called "Inside Alaska." I've seen most of the state.

But the station was sold, the show cancelled. My years at WAAY were some of the best in my career so far. While there may have been some editorial differences, I always felt I was treated fairly and professionally, which is more than I can say for some of the other places I've worked. I hope all is well with you and your family.

Regards,

Kirk Chaisson

[*Editor's note: I asked Kirk did he remember the tornado of '89 and this is his response.*]

Interesting response. In fact, I had just arrived in Huntsville when the twister hit, but that is not very high on my list of recollections, because I was new there and didn't do much reporting on that particular story.

What I remember most is the several hidden camera and other undercover investigations that Jeff Raker and I did, in which both he and I literally put our lives on the line at least a half a dozen times, whether it was stories about the Ku Klux Klan, murderers on Sand Mountain near Scottsboro, or any other number of award-winning investigations. Don and Heather anchored, and we did some pretty good T.V., certainly just a small part of the great T.V. you facilitated over the years.

Lisa Waugh – "WAAY Too Late" (W2L) producer.

Hi M.D.

I hope you and your family are doing well. I must apologize that I am in the middle of post-production on a film and I'm scattered, smothered and covered to use an old Waffle House term.

I can give you a quick story here and let you use it however you wish.

We used to commando shoot W2L. I don't mean we did it in the buff (although Kip kept suggesting it) but we would shoot all over town in the middle of the night while the proper authorities and business owners were safely in bed.

One night we were shooting a weird little sketch in a parking garage. The scenario was that Kip and Scott couldn't get out of this garage (for the record - this was way before that freaking Seinfeld episode) and they run across all of these bizarre characters.

They happen upon a hardcore feminist possibly lesbian compulsive fast walker on her nightly jaunt in the parking garage. They ask her for assistance and she goes zoo on them.

Now we didn't rehearse this scene (hell, we barely wrote anything down on paper) and I was supposed to beat them up then pummel Scott with a baseball bat.

I tried to be careful. I really did. Scott's scream could be heard all the way to city hall. I'm surprised the cops didn't show up.

I really whacked his knee good. I think we have this on tape somewhere.

I loved working on "WAAY Too Late." I made some life-long friends, dated a couple of those boys and it prepared me for the career I'm ass-deep in now. Hollywood ain't got nothing on no budget, middle of the night dead on your feet shoots, and eight viewers.

Oh, and I also popped Scott in the mouth that night, too.

Sorry, Scott. I love you, man. [*Editor's note, here's a further comment about Scott from Lisa]* Scott is Scott McCauley. Funny thing about Scotty is that we put him in drag so many times and he never complained. He played my mother once and freaked me out by holding up his foot and asking in a raspy voice, "Does this look infected to you." I still have nightmares.

I'm off to sit in a dark room with an editor, smoking too much and nitpicking the audio until its right. Just like those nights at Channel 31 with my buddies of WAAY Too Late.

Well, M.D. I hope that works for you.

Don Phelps – Anchor-Reporter from 1985 to 2002 & God

GOD LIVES AT WAAY-TV

It was the last commercial break in the 6:00 News. Heather Burns and I were waiting to do the last couple of stories when the door, next to the set, flung open. There stood a wild-eyed young man in his 20's. He was wearing a sweat drenched white t-shirt, blue jeans, and no shoes.

Before Heather and I can react, a slender hand grabs the man by the back of his shirt and drags him out of the studio. The door slammed just as the commercial break ended. We heard some scuffling and bumping at the door while we finished the last block of news, but the man never made back into the studio before we finished the newscast.

We didn't know who the man was but we knew the hand belonged to Yodie Lockhart. Soon the lobby was crawling with police. We could see bits and pieces of what was going on, through the door's window, as we were cutting the teases for the 10:00 newscast.

Just as we were getting ready to leave the set and find out what in the world was going on a police officer came into the studio and asked us to keep our seats.

He explained the man was a patient on the mental health ward at Huntsville Hospital. From his room he could see the cross on Monte Sano Mountain and the flashing lights on the T-V tower. The young man escaped from the hospital and ran all the way up the mountain in his stocking feet. He knew if he could get on channel 31 he could talk directly to God. The officer said the young man was very agitated and refused to leave until he talked to God.

The officer feared they were going to have to hurt the young man to get him out of the station. The officer then asked if we could put him on T.V. The officer was convinced if the guy had his say he would leave with them quietly.

There was no way we were going to put this guy on live T.V.!!! So, I suggested we take the monitors off the "On-air" signal and put a program feed into them.

Heather and I faked the ending of the newscast, tossed to this guy who was in the studio with us now. He had his say and was now a happy camper. The studio crew and Tim Clemmons and the production crew played along with it. They took a bump shot from the set and even rolled the close to the newscast. It was so real one of the cops asked.... "You really didn't put him on T.V., did you?"

Two weeks later the guy shows up at the station again in his stocking feet. This time the police told him he would have to communicate with God the normal way, by prayer.

Mary Beaton Morgan – Morning Show, 10 Years – Sweet

Thanks for calling. It was good talking to you. Henry and I are looking forward to the reunion party.

It doesn't seem possible, but I worked for 31 for almost ten years and had ten different partners on "Coffee Break"," Morning WAAY", "Mid WAAY ", and then "Morning WAAY" again. My first, and probably longest partner, was Johnny Evans.

Johnny was great to work with and we had a great time together. I almost quit after the first show though. Johnny had scheduled a guy from a small traveling circus to come on Coffee Break. (This was a time when Huntsville was much smaller and it was pretty hard to get different guests each day.) The man from the circus brought a not very well-trained bear. As soon as he got on the set with all the lights, the bear went on a rampage. He climbed on the desk and we had visions of it crashing down and crushing our legs. Johnny kept his cool though. I remember him saying, "cut that out - this is my show." I don't remember how we got him out, but after a break we were able to go on with the rest of the show.

The people at 31 were like my extended family and we went through lots together. My husband died during my time on the show and my daughter was ill during that time.

We left Huntsville in 1978 and moved to Anniston. We were there for four years and then moved to Decatur. I now own Morgan Price Candy Co. and stay very busy with that. My company started out as a small mail order company at Christmas but now have about 10 or twelve employees depending on time of year. Can't think of any other business that would be more fun. Everyone is always in a good mood when they are buying candy.

From TV Guide, May 16-22, 1970

Al Whitaker – Reporter, Anchor, Shoals Bureau – Wild Stories

There's no single favorite memory, no defining moment I can draw upon to represent my time at channel 31. But there are a few moments that stand out, like the night Kevin Athow and I launched bottle rockets at another employee from the roof of the station…Another night, he laid some "Snap-n-Pops" on the floor of the restaurant where we'd stopped for dinner. A waitress stepped on one or two and within a few minutes, workers were taking the ceiling tiles down looking for the source of the popping noise…The early morning live shots with Rob Ballard, including the morning he had me perched on the railroad crossing sign high over Church Street at the depot. It really pisses off the railroad folks when you do that…Jeremy Woosley's "By the WAAY-side List"…Cactus Gay and the Christmas parties…Flying through the eye of Hurricane Bertha with Gary Dobbs…Learning in the middle of an interview at city hall Athow's wireless mic was on the same frequency as the wireless mic in the Huntsville City Council chambers, it took us several moments to figure out why so many people were running towards us waving their arms…

The night Decatur Police threatened to arrest me for aiming the laser pointer towards city hall (I didn't know it was the magistrate standing outside, she thought someone was pointing a gun at her)…The day Adrian announced it was so clear outside he could see the back of his own head…Friday Night Football and the food cams…And the night the coach stepped into my shot of a running back making a touchdown. That night on the air I tried to say, "Folks, don't adjust your set, that's the coach's shirt." But for weeks after that, several times a day, people would come up to me and say, "Hey Al, don't adjust your shirt!" Pierre Kimsey was in the bureau that night, and if you play the tape back and listen carefully you can hear him laughing to the point he was struggling for air.

…The day I renewed my contract and drove back to Florence to cover a tornado strike. Later that evening, Alex Lynch and I would return to Huntsville and spend the night doing live shots along Airport Road, where another tornado had hit. The temperatures plummeted below freezing and after I'd shivered through several miserable hours of live shots, a man walked up with two heavy work coats he'd brought from home saying he had been watching and could tell how cold we were. He'd walked several blocks from where he had to leave his car to find us. We only saw the guy a few seconds, just long enough for it to sink in why he was there and just as abruptly as he arrived, he turned and walked away. I have no idea who he was, but every time I think of that horrible night, I see his face, and Alex's as we put those jackets on…

Speaking of Kimsey, he invaded the Decatur Bureau for what seemed like months, searching daily for one of those off-the-wall feature stories…His mind seemed to be elsewhere much of the time. You knew when you had his attention, and there was no point talking to him if you didn't.

He and photographer Tommy Gamble produced a story once that had Pierre trying to deliver a news tape to the bureau for me to feed to Huntsville for the next newscast only an hour or so away. We relocated most of my office from the sixth floor of the AmSouth Bank Building to the third, where there was a balcony.

At about the time I'm on the phone explaining to Mark McGee, "If Pierre doesn't get here within the next few minutes, that story's not going to make it," Pierre is standing outside the locked bank building looking nervously between his wristwatch and up to my office window when he spies the answer to his dilemma: The janitor's cart, complete with two toilet plungers he'll use to scale the outside of the building up to my sixth floor window. It was classic Kimsey and the finished piece was not only hilarious, but the sequence of him scaling the building with those two plungers was absolutely believable! That was key to Kimsey's genius; he would lose himself in these ideas and solve every minute detail long before the camera was turned-on.

To achieve the shot of Kimsey pulling himself up the side of the building, he and Gamble went across the street to Decatur City Hall where the tone and texture of the sidewalks resembled that of the bank building's exterior. Gamble positioned his camera on an upper floor patio of city hall shooting almost straight down upon Pierre, who simply crawled up and down the sidewalk with the two plungers as though he were pulling himself along. When that shot was blended with a static shot of the bank building, it really did look like Pierre was inching his way up using nothing more than those plungers.

Pierre was apparently too caught-up in the details of the shoot to consider talking with the folks in City Hall, including the police, about his plans. So caught-up, that when the man who first saw Pierre crawling along the sidewalk that way asked if he was alright, Pierre barely looked up long enough to mumble part of an explanation he hoped would send this good Samaritan on his way, something about "scaling the building" the man told police. Within moments, Pierre had crawled up on two highly polished pair of black leather shoes. This time he stopped and raised his head, eventually looking up and directly into the faces of two Decatur Police Officers.

"Sir, I know you probably have a logical explanation," one of them started, but Pierre was already speaking to them in incomplete sentences, trying perhaps to draw the officers into the moment in hopes they may appreciate the creativity of the plan, or something like that. And it might have worked except Tommy had stepped back to light a cigarette. So, when Pierre came to that part of the explanation, and there was no photographer up on the patio, he and the plungers were promptly led away.

The next day, as I went from one office to the next to apologize for the disturbance, I was told word of the "maniac on the sidewalk" had spread through the building quicker than fire with one office calling the next, and so on. Even before the police got to Pierre, work on all 12 floors at city hall had stopped as employees pressed against windows on that side of the building for a view. For a few brief moments, they honestly believed he was going to try to scale the outside of their building with nothing but a pair of plungers…

Jeff Rosado, Director-switcher & Live Shot Gone Bad

In an earlier email, MDS asked about a blooper tape where a reporter was on a live shot and she didn't know she was on the air and something happened and she loudly and clearly said, "OH, SH#$!!!"

That would be Susan Holley. -- I was on duty when that occurred, we broke in during Rosie O' Donnell, so we had a lot of people watching. News was trying to get a jump on the competition and I remember Jeremy Woosley (one of our on-duty engineers at the time) going, "Take it, take it, take it!" -- Although hesitant to do so feeling we were moving too fast, what could I do? Cross faded from that goofy special report open with cheesy music and all to her...and of course, the rest is history. *[There was a fire at a lumber yard Susan was covering, and just as she went on, there was a car crash that made her yell.]*

What made it even funnier to me is that after I faded to black quicker than I ever had in my life (and I had gone completely numb at this point), where else could I go but back to "that goofy special report open with cheesy music and all." Then after a few seconds more, Jeremy more calmly said, "OK, go ahead, Jeff," to which I tersely replied, "Are you sure?"

And the icing on the cake was that priceless look on Susan's face when we returned to her (and if you listen hard enough, you can hear Kevin Athow saying, "Go.") -- I think at least 10 copies were made in a span of a half hour alone following; that was when the master control ops had to record all the special reports and weather coverage in full. Poor girl. Last I heard she was in Colorado doing tv work up there.

Well, I think I just contributed to your alumni page. -- Jeffrey

Evelyn Starnes (Pelfrey) – Promotion Manager '77

[Editor's note: Evelyn was promotion manager in 1977 when we switched to ABC. She, Dan Whitsett and I traveled to L.A. to attend the promotion manager's meeting to get hyped up and learn how to promote ABC. We got it in spades. I think the "Still The One" theme this particular year was the best ABC ever did. I may be partial but listen to all the music of these years and see what you think.]

Well, if it isn't the boss who gave me my first break in media right out of college! Yep, you found me. I checked out the WAAY alumni Web site and got a blast from the past! I still use Starnes as a middle name, but my by-line is now Evelyn S. Pelfrey. I'm a newspaper reporter for the daily paper in Southeast Alabama, The Dothan Eagle. You probably remember I went to WSGN Radio News in Birmingham after I left WAAY. From there, I went to WAPI AM/FM and then got a big break and went to NBC Radio News in New York in 1980. Eighteen months later, I went to NBC owned-and-operated WRC Radio in Washington DC, and then on to the NBC Washington Bureau before becoming News Manager at Mutual Broadcasting System in Arlington.

I got out of broadcasting for a decade and moved back to Alabama when my husband's mother passed away and his father couldn't live alone. After a few years as a newspaper reporter for the Barbour County weekly, The Clayton Record, I moved on to The Dothan Eagle early last year. And you gave me my start! Thanks!

Thanks for finding me! Evelyn

Don Roden – Engineer & Tower Climber, Bad Weather in 70s

April 3, 1974 (Night of the Killer Tornadoes):

Carl Sampieri and I were sitting in the floor with three RCA AVQ-10 radar antenna assemblies...none of which worked. The slip rings that transferred power through a rotating joint were burned due to the high humidity and rains of the previous month. The elevation control was intermittent, or worse, would point straight down as the slip rings lost contact. "IF" the antennas worked, there was a more serious problem...the indoor display was being "killed" by ground return.

Carl was hired as our "resident radar expert."

Cactus reasoned that:

(1) Carl had just returned from Viet Nam after a tour with the Air Force.
(2) The Air Force used F-4 Phantom fighter planes.
(3) F-4 Phantoms had radars.
(4) All radars were similar.

Sounded good to me. I was glad to have some help.

Trouble was...this was the first radar Carl had worked on; we were both lost; and bad weather was on the way.

We flipped coins to see who would climb the tower and put one of the antennas back in the radar dome...I lost.

I climbed the tower in a light rain shower and was glad to get inside the dome for shelter. I bolted the antenna to the brackets and used some rope to prevent the elevation from falling down. By 4 p.m., we had the radar making a very weak picture. Something wasn't working right yet. We decided that "some radar returns" were better than "no radar returns," and that we should leave everything alone until after the storm passed.

The storms that hit later that night were dumping so much rain that a good radar would have been overwhelmed. Our "insensitive" radar, on the other hand, showed only the most intense hook echoes. One of the storms took out the NWS radar at the Jetport, and 31 was the only station running live radar that night. The other stations were watching 31 for their weather coverage.

Don Roden

Scott McCauley - Production & Photographer – Bombing Iraq

Favorite Quotes from My Generation:

On a 10:00 Friday night show, Heather Burns, Don Phelps and Gary Dobbs were wrapping up. Gary plugged the Guntersville Pork Festival occurring the following day. Without missing a beat, Heather innocently said, "If they're not fishin' in Guntersville, they're porkin'."

Another case of the Friday night sillies, Heather began the kicker with a question, "What do you have when you..." and listed items like several hundred pounds of corn meal, ham, gallons of iced tea. She concluded the list with 500lbs of dried white beans. Don replied, "Probably enough natural gas to heat the Tennessee Valley for an entire year. But if you're in Florence, you have Bean Day."

Other Events of Note:

About a year after Operation Desert Storm had concluded, ABC had sent down a refeed of its special report break-in it did the night it all started with Peter Jennings tossing to spectacular night video of Bagdad being bombed. Well, the switcher on duty did not see the file video slate prior to the start of the feed. When he saw ABC Special Report slate, he put it on the air. And we proceeded to scare the Tennessee Valley by making them think we were bombing Iraq again. Phone calls poured in asking if we were attacking Iraq. After the mistake was realized and regular programming continued, ABC sent down an alert in big white letters telling its affiliates that its previous feed was file video only and was not for air. Apparently, many other switchers in many other markets were also quick on the switching trigger to take the ABC Special Report.

We were shooting a Christmas feature for Pierre Kimsey at Parkway City Mall. The premise of the scene was to show the frenzy associated with Christmas shopping. So, we were posing as shoppers and acting like we were choking each other over a parking space. As we were acting the scene on camera, I got to do the choking. Suddenly I felt to firm hands on my shoulders, and I was pulled back and separated from my chokee by mall security. They had received a tip from a shopper that there was a fight going on in the parking lot. Evidently, security did not see the large camera and tripod sitting in the back of a pickup truck shooting the entire thing.

Linda Pope - Going to Graceland – What Not to Do

"WAAY Too Early" remote at Graceland and the 3 major No-Nos:

1) We were told by Graceland's marketing director, under no condition, NOT to sit on any furniture in Graceland. Pierre Kimsey somehow did not get this information, or shall I say did not absorbed it. He was found before the show started, kicked back on a VERY white and formal couch in the "Blue" room with his script spread out in front of him on the King's coffee table. We were coming and going setting up lights and cameras and whispering to Pierre to get up. He was very much in deep thought over his script and was not realizing his error. I think Gary Dobbs had to finally poke him on his shoulder and ask him to follow him out into the foyer and re-relay the direct order of staying off the furniture.

2) Towards the end of this show Toni Lowery was outside by Elvis's grave, I believe, thanking all the sponsors, including the Memphis Holiday Inn and Luby's Cafeteria for lodging and feeding us during our time in Memphis. The marketing director, who had been

conducting a tour on camera for our audience in the Tennessee Valley was inside the house with Gary off-camera and could hear Toni's plugs. This man, who is already uptight about us being in Graceland since midnight, became visibly upset at hearing this. **It turns out that Graceland has never in its history given permission to any one or any corporation to use Graceland as part of a sales pitch**. They were turning down Japanese Corporations that were calling on a daily basis with multi-million dollar offers for permission to use Graceland in their advertising. (He is telling Gary this with clenched fists as Toni is plugging away outside.) After cutting away from Lowery to the indoor shot, Gary strays from the script (this was his time to plug sponsors) and talked to the marketing director about the house, again, but rightfully so. Toni is hearing this outside and is upset that Gary isn't doing his planned sponsor plug. She was quickly informed about the faux pas.

3) To add insult to injury: After the show is over, the equipment is packed away and apologies are made. Shawn Jarrett, Toni Harper and I were packing up the production van when the 31 Live truck rolls out of the circular driveway and leaves us standing (with the marketing director in close proximity) by a very LARGE patch of BLACK grass by the front entrance...the area where the exhaust pipe from the 31 Live truck had burned the grass during the show.

What a hunk of burning love we left.

Kip Cole & "WAAY Too Late" – The Real Story

WAAY TOO LATE VERSUS THE ADDY AWARDS

"WAAY Too Late" first aired in February 1991. For those who aren't familiar with W2L, it began as a way to host movies that were to be shown late on Saturday nights. Really, really BAD movies, for the most part. Program Director Debi Benson had the brilliant (some might call it unfortunate) idea that this might be a good opportunity for some station employees who normally labored behind the cameras to showcase some of their hidden creative urges. Seems the poor woman was under the delusion that there was gold to be mined from some of us poor schleps who pushed the buttons and ran the cameras for our newscasts.

At first there was supposed to be only one host (or pair of hosts) for the entire show. But after a bunch of us turned in audition tapes, the dubious decision was made to allow several of us to act as hosts on a rotating basis. Originally there were five sets of hosts: Kip Cole & Scott McCauley, Chris Homsley & Stan Williams, Jeff Rosado, Priscilla

Cowing (who brought along Lisa Waugh as her producer) and Renard Vandross (the self-described host with the most). Later we dropped Priscilla and Renard, added Lisa and Laura Bailey as a host duo and a bit later Mark Bowling, Shawn Jarrett and Tom Barker. At one point, Keith Matthews was in the group briefly, then Mike Best and Pierre Kimsey, a one-man film school, who taught us more in a few months about storytelling and editing than any of us ever learned in school.

But for most of the show's run it was the core group of ten of us: Myself, Scott, Tom, Chris, Stan, Lisa, Laura, Jeff, Mark and Shawn. We all became great friends, a family if you will…much like the Addams Family, only more dysfunctional. In spite of the "rotating host" format, most of us were involved in just about all of the shows, even if we weren't hosting that week, acting in or shooting or editing each other's segments. We became a true ensemble in every sense of the word.

WAAY Too Late had no budget and no scheduled production time. All of us had regular jobs at the station, some working news, some WAAY Too Early, a few in commercial production, so we had to work on W2L on our off-hours, which was usually nights and weekends. And by nights, I mean ALL night…I can't count the times a group of us would still be at the station shooting or editing when the WAAY Too Early crew came dragging in at 5 a.m. Many times, we would still be working on the show (which aired at 1 a.m. on Saturday nights) AS IT WAS AIRING…i.e. editing one segment of the show while a segment of the movie was on the air. I can still remember the frantic pages from Jeff R. (who switched the show):

"ONE MINUTE OUT!" Eject the tape. Hand off.

Sprint down the hall, through the studio (as other members of the crew held doors open). Slam the tape in the machine. Hit "PLAY" just as the segment hit air. The rush of adrenaline followed by cardiac arrest. The movie "Broadcast News" had nothin' on us.

When we started WAAY Too Late, none of us knew what the hell we were doing. We all had certain skills that we developed in our regular station jobs, directing, editing, shooting, etc., but none of us had much experience with linear storytelling. And in those early days, I gotta admit, it showed. The decision to air W2L was a brave (some may call it foolhardy) one on MD's part. To allow a bunch of kids access to expensive broadcast equipment, shooting what amounted to experimental shorts aimed at an extremely eclectic audience (mainly ourselves), and then to give us AIRTIME...well, no matter how ill-advised, it was a ballsy move.

MD once referred to the show as "an interesting experiment," but for us it was, pure and simple, film school, only without actual instructors and high tuition. The early W2L shows were pretty rough, although there was a glimmer there of something...I don't presume to call it talent, but you could sense the flow of creative juices (which Junior Morris and John Langford had to repeatedly mop up), no matter how far the skit fell from the original concept or how poor the execution. But we did get better. We learned as we went, and our later shows, if falling short of great television, at least did become something that we could be proud of (or at least not be ashamed of telling our children about). The original concept of WAAY Too Late, as laid out by Debi, was to "host" these movies (did I mention how abysmally BAD most of them were?) offering information about the films in a clever and creative manner. Well, we started out that way, but soon grew tired of trying to make movies that made Hammer films look like they were directed by Fellini seem interesting. I mean, there's only so much you can say about a movie called "Die Monster, Die." So, we eventually stopped referencing the movies altogether, just doing whatever skits our feverish little imaginations could conceive of. And looking back on some of those shows, the remarkable thing is that no drugs were involved. We should have made that our disclaimer. After two years or so of being a weekly show, W2L evolved into a half-hour, stand-alone sketch comedy show that we aired once a month (or as often as we could get one put together). In those half-hour shows, we had finally come into our own as far as our creative and technical abilities went. Some of the sketches we did then were truly funny and well done (and some...not so much). By then we had Pierre on board to guide us, teaching us comic timing, editing and the importance of brevity. We only did five or six of those shows before we all began to go our separate ways, becoming busy with other projects, but they were the highlight of my time at WAAY-TV, if not all of us. As I said, we became fast friends, most of us still in contact today, wherever we are, be it L.A., Memphis, Rome, GA or Huntsville. It still amazes me that at one brief point in time, at a small TV station in Alabama there could exist that many people possessed of the same mind, the same sick, twisted sense of humor and sensibilities. And, as I've always been fond of saying, it was much cheaper than therapy. So that's the story of WAAY Too Late in a nutshell. But that's not what this story is about. This is about the one, single shining moment in the brief life of W2L, the one night where all of our hard work, long hours and incessant ingestion of nicotine and caffeine finally paid off. The night we were vindicated. The night of the 1992 Addy Awards.

From the very beginning working on WAAY Too Late was an ongoing battle. We were the red-headed stepchild of the station, a rogue group that didn't seem to fit into any previously defined category at Channel 31.

The Sales Department hated us, presumably because the show wasn't prominent enough or aired too late to bring in enough revenue to warrant going after advertisers. The Engineering Department didn't like the fact that we used station resources to produce something they deemed to be frivolous. The News Department tolerated us, containing both supporters and detractors to the efforts of W2L. We got our share of complaints from News; one anchor complained that in the odd hours we spent working on the show, the station was filled with "unfamiliar faces, coming and going at their leisure" (which we promptly ridiculed in one of our skits). But Don, Heather and Gary usually seemed eager to humiliate themselves for us whenever we asked them to be in a sketch.

But as I said, WAAY Too Late was a constant battle. It seemed like we became the station scapegoats, being blamed for anything that might be amiss in the building. From damaged equipment to holes in the walls to bruised egos, the finger of blame inevitably would be pointed in our direction regardless of the crime or the actual perpetrator. I can't count the times I was dragged into Lew's office to account for some incident that had occurred the night before. And to be perfectly honest, there were probably many things that were our fault (and quite a few that will never be uncovered), but to be involved in every damn thing that happened in that building…well, we had neither the time nor the energy to perpetrate all those heinous acts of criminality.

But we did have our supporters. Debi Benson served as the executive producer and Lew Koch as our supervisor, since most of us were in the production department. They stood behind us through all the ups and downs of the show. The numerous times that we were slammed in Department Head Meetings, our collective heads being called for on a platter by other Dept. Heads, Lew would always come to bat for us, defending us to the death. He was Lancelot to our Guinevere. Er...or something.

And of course, MD. Every time someone from sales or engineering called for W2L to be yanked, MD would stand his ground, allowing us the freedom and resources to continue doing this little TV show. Oh, there were the occasions when I'm sure W2L threatened to cause his ulcers to start bleeding, getting the occasional voicemail from some irate caller whose delicate sensibilities we offended, some poor insomniac who tuned us in by mistake when expecting to get some sort of spiritual relief or psyche massage from Jerry Falwell or Robert Tipton, only to be assaulted by visions of Scott in drag or Lethe eating a Vienna Sausage. Even in those awkward moment's MD stood fast behind the show and for that I think I can speak for the entire crew by saying thanks.

By the end of 1991 WAAY Too Late had been on the air for almost a year. In that time, we had come a long way, learning a lot, honing our craft and eventually producing segments that left the realm of "bad," surpassing "mediocre" and becoming, in some instances what could be considered "pretty damn good." And, we hoped, remotely funny.

So, it was with great enthusiasm that we produced a highlight tape to submit to the local Addy Awards that year, in the Potpourri category, hoping against hope that we would at least score a nomination, if not actually win an award.

Shortly after submitting our entry, we were contacted by some folks from the Addys. They liked our tape so much they wanted us to produce segments that introduced the various categories on the presentation tape that would air at the awards ceremony (a hoity toity affair held at the Hilton or someplace that brought together for one night everybody who thought they were somebody in Huntsville broadcasting). It wouldn't be too big of a deal, they said; we just produce, shoot and edit the segments and Channel 48 would take them and put them into the two-hour presentation tape that they were in charge of editing that year.

No problem. OK, well, a couple of problems.

The first problem was the "theme" of that year's Addys. Each year the Addy Awards has a theme...one year it had something to do with cavemen and the discovery of fire or something...one year it was cinema related...anyway, you get the picture. The "theme" of the 1992 Addys was "The Addy Dive-In." No, that isn't a typo. It really was called "Addy Dive-In."

Some brilliant bastard in the advertising community came up with a concept that combined two subjects: '50's drive-in diners and swimming.

Yeah, I know...two subjects that immediately go hand-in-hand in my mind.

But the thing is, other brilliant bastards in the advertising community thought this was a great idea. While the concept admittedly made for a cool looking poster (some Avant Gard piece of artwork that had a guy in a swimming cap coming off a diving board attached to a diner), it presented a bit of a creative challenge for video skits. But somehow we managed to make it work, shooting segments at the Natatorium, Rocket City Diner and a drive-in restaurant. What we came up with, while maybe not comedy genius, was at least fairly clever and creative.

And of course, we managed to have Scott in drag, playing a waitress from hell.

So, we got our skits done and were ready to take them to Channel 48 to be edited into the presentation. Since we had worked so hard on the Addys thus far, we were reluctant to just hand the tapes over to some unknown editor. Our stipulation was that some of us would work with the 48 editor in their edit suite to make sure the skits were used the way we intended. WAFF's Production Manager, who was an incredibly nice guy agreed to this, so we set up a couple of overnight editing sessions for us to come over to 48 to work on the show.

Then came problem number two. It seems that Channel 48's GM wasn't too keen on having a group from Channel 31 hanging around his station after hours. At that time, WAFF was the lowest rated newscast in the market. I guess there was some secret to being ranked number three that he was loathe to let spies get their hands on, so he refused to let us in the building. 48's Production Mgr. called me at home one morning to give me this news, sounding a bit embarrassed at the situation.

So we decided, screw 48, we'll edit the thing ourselves. Lew graciously let us use Morrison Martin to help edit the presentation and we spent the next week or so working all night every night to finish it up. Most of us provided the voice overs announcing the nominees and winners, usually in some bizarre character voice we had come up with in the middle of the night while punchy from lack of sleep or wired on caffeine. It was great fun, and as was usual for W2L, we finished up about 8:00 on Saturday morning, the day of the Addys. And, we hoped, making Channel 48 look like a bunch of assholes.

Now here's the thing. The winners of the Addys are supposed to be a closely guarded secret on par with the identity of "Deep Throat" or who really killed JFK, to be revealed ONLY on the night of the actual awards. Well, since we were producing the entire presentation and had to know who the winners were, the Addy people reluctantly informed us that WAAY Too Late had won an Addy in our category. We were, to say the least, quite pleased to learn this and would later use this bit of information to our own insidious advantage.

OK, it's Addy night, sometime in January 1992, almost a year since W2L began its checkered career. We had our own table, a rowdy, unkempt bunch in a room full of stuffed shirts, so brimming with self-importance that you'd have thought it was a meeting of the U.N. But we were the rebels, remember. The red headed stepchildren. We were loud. We were obnoxious. We booed, hissed, belched and hooted. Shawn and Mark swiped an entire roll of drink coupons providing the means for most of our group to get completely shitfaced.

I remember Mark and his then girlfriend Leah (soon to become his wife, then ex-wife) drunkenly shouting some very inappropriate things during the playing of a rather somber and reverent PSA spot.

I remember us having to practically pull Lisa off some hapless video projectionist who mistakenly turned off the projector about thirty seconds too early, cutting off one final joke at the end of the presentation. She had murder in her eyes, booze on her lips and was spewing a stream of obscenities that would have made John Ashcroft's head explode. That poor sap will never know how close he came to death that night. And I remember the surprise of the night, when it was announced that we had won a special award, one created just for us, a "Judges Award for Creativity and Imagination." We almost missed the announcement because of all the drunken shouting coming from our own table. Somehow, I managed to hear our name called and staggered to the podium to accept the award, a rather impressive looking gold star-shaped statue with WAAY Too Late emblazoned across the front.

Now, here's the other thing. The way Addy Awards work is like this: whatever project wins in its category receives a single Addy Award the night of the ceremony; whoever was involved in that project can then order duplicate statues after the fact. That night we had girls dressed in waitress costumes deliver the Addys to the winners' tables on a large tray (in conjunction with the diner theme; some of us guys wanted them to be in bathing suits, but that idea didn't fly with the Addy folks for some reason).

Well, since we knew ahead of time that we had won in the Potpourri category, we went ahead and ordered Addy statues for every member of the group. The Addy people, appreciative of all the work we'd done on the program, didn't seem to have a problem with this request. And at the time we didn't think that this might overstep the boundaries of some unwritten rules of etiquette in the realm of advertising awards. We were just proud to have something to show for our year of toil on WAAY Too Late.

So, toward the end of the night, when our name was announced, this table of drunken, unruly kids who had gotten progressively more and more irritating throughout the course of the evening, received not one but TEN Addy statues brought to our table by two of the "waitresses."

It wasn't until much later that we learned how much this one innocent act pissed off so many of the muckety mucks in the Huntsville advertising world.

Not that we really cared, mind you.

Yes, that night we were obnoxious. We were loud. We were irritating. Some were drunk and disorderly. But that night we became something we had never been before, as a group. We were winners.

After a year of constant battles and threats, after a year of being sleep deprived, living off cigarettes and Diet Mt. Dew, we were finally, irrevocably vindicated. WAAY Too Late might not have ever developed a large audience, but that night, most of Huntsville broadcasting and print knew our name (albeit most likely with an attached epithet).

I'm sure that all of us will remember that night. At least those not too intoxicated. And in the years since there have been other awards, other accolades. Some of us have won Emmys, some have written for network television shows and feature films, most have gone on to bigger things. But for me, at least, no matter how much I accomplish, what awards I may win or amount of recognition I receive…they will forever be measured against that night.

The Huntsville Addy Awards, 1992. Kip Cole Memphis, TN July 20, 2003

Michael Sheffer, Engineering, Impossible Live Shots

The one thing you'll hear time and again from people who were part of the WAAY family is the amount of leeway we were granted in any number of endeavors. We allowed to express, explore, and drive ourselves to our own limits in finding out what we were capable of - even if we really had no clue at the time. It was a breeding ground for raw talent to grow in a variety of differing fields.

Many folks from the production side of the station have brilliantly commented on the freedom of expression they were allowed, but I'd also like to recognize the WAAY Engineering Department as another shining star in the 31 constellation. While there are innumerable stories about "Cactus' Boys" which could fill volumes, I'll play it safe, and focus on WAAY's unique ability in engineering live field remotes.

No matter what it took, no matter how far we had to go in order to bring back the feed; from double hopping a microwave signal from Sand Mountain–with the repeater perched precariously atop a 40 foot billboard, to setting up a full blown, 5 camera live news program on little more than an hour's notice, our little group succeeded in an area that most top market stations fear to tread. If it was "remotely" possible, we were always able to find a WAAY. There was nothing more rewarding than knowing that our competition would simply scratch their heads and wonder; "How'd they do that?" when we hit the air.

WAAY was the very first station in the market, and possibly all of Alabama, to produce a live, satellite fed newscast. Under the direction and designs of John Hain, we developed any number of innovations, from the Piggy Pack camera remote interface, to what may arguably have been the very first "in car camera" for our racing coverage at the Huntsville Speedway. In order to facilitate reliable communications, we developed our own pro channel receiver, so that studio commands and cues could be heard in the field without having to rely on telephone or (at the time) expensive cell phone links. We grew from the old LIVE EYE; a massive Ford Econoline Stretch Van that drove more like a rowboat in rough seas to a full-blown mobile production vehicle that is still in use (last I heard). A photo of the new truck even graced the front page of TV Technology Magazine. Our sports coverage, especially motorsports, won accolades from the Alabama Associated Press three times in 1992. Think about it—a Statewide news organization recognizing an engineering department...Only at WAAY could it have happened.

Donny Seymour and I would roll out the truck for any number of different projects. Whether it was Johnny Evans' "A Country Affair" music program, or Jamie Cooper and Gary Dobbs hamming it up at some automobile dealership at 4 in the morning for "W2E", over the years we had become the most efficient remote team in the region. We had a huge amount of fun and developed the skills which we use to this day.

It's a unique experience when many different and talented individuals come together and jell into such a productive and powerful group. Nothing got in our WAAY–either we simply didn't know that "It's impossible" or we didn't care; somehow, the job got done. With all of the hard work, ego clashes, fried equipment, frayed nerves, blown fuses, and everything else that can go according to Murphy's Law, I'll always value my years as a WAAY family member, and am very proud to have been one of "Cactus' Boys".

Oralie Tyson - 1989 Airport Road Tornado

Reflections of my years at WAAY - TV bring back many humorous anecdotes, but none is appropriate for public disclosure! However, the one continuing memory I do have, that can be documented, is the sense of "family" we all enjoyed--(yes, families DO have internal disagreements, so we qualified!)

This sense of family was most clearly illustrated during the BIG tornado that went down Airport Rd, killing more than 20 people. Everyone in the station pulled together to help each other, the community, and keep the public informed in a time of crisis. EVERYONE—producers, admin folks, reporters, salespeople, engineers, etc. rose to the occasion. People went into the storm and returned drenched from head to foot, only to go out once again into the rain and snow to what resembled a war zone on Airport Rd., dedicated to keeping the shocked Tennessee Valley informed and calm.

Through the years, those of us lucky enough to work at WAAY -TV came to appreciate the "family spirit" of the station–along with the sense of encouragement to grow and excel that was Ch 31's style of management.

What other company would have a "Grape" or "WAAY Too Late"? or encourage the employees to give feedback on so many important issues, ranging from which syndicated shows to buy, to specifics on how to better run the station.

And now as the years go by, I appreciate even more those extraordinary years spent at WAAY TV–an incomparable and unique experience marked by many highs (remember when our news numbers EXCEEDED the combined numbers of 19 and 48, and we celebrated with ice cream cups!) WAAY-TV made a difference in many of our lives. Thanks for a great ride, MD!

Redge Swing - Local Sales Manager, "No Bull" Inspiration

From a Sales Point of View

Coffee Break, Bull and Magic I probably learned more about TV sales listening to my office mate of over 20 years work the phones than from Willie Gale, Earl Nightingale, Dale Carnage and Zig Ziglar combined. For sure, I was a quick study under the guidance of Johnny Evans when it came to trade-outs, live plugs and promotional mentions. I wondered why Johnny always got so many boxes in the mail until he started sharing the "promotional items." As I recall, several departments had folks very adapt in these areas. I still believe we may have been contributing factors in the closing of the Diplomat and Twickingham Station.

BS took on a whole new meaning when one of Johnny's clients showed up one morning for Coffee Break with a 4000 lb. prized bull named Rolo II. The client was very put out that Rolo couldn't come in the studio for his appearance. Seems like I recall Coffee Break co-hostess Mary Beaton reply we already had enough bulls@#t in the station. Not true, Mary is a super sweet southern lady who would never say such a thing. Thanks to trucking camera, a strong cameraman and stretched cables, Rolo II appeared live from our parking lot. This opened up a whole new outdoor studio.

In addition to a ventriloquist via Tex and Grandfather clock (one could get an argument from Tex as to who was what), Johnny considered himself a quite a Magician. There is an old saying "it takes one to know one." This didn't hold true for a slight of hand trick one morning on Coffee Break. A local flea market vendor was scheduled to draw live for the winner of a used car folks had been registering for at his flea market booth. We had been promoting the registration and drawing on our air. After the drawing and the announcement of the lucky winner, our switchboard received an anonymous call identifying the winner as a close friend of the vendor. It was suggested we review the tape of the show. Sure enough, upon further review it was very clear the vendor had not so neatly palmed the winning ticket. A redraw was done later without the magic.

Early Sales Trip

Some salespeople were slower than others when it came to trades. Dan, Johnny, Dean and I were on a sales trip to Ft. Walton. For what reason, if there was one, I don't recall. We were picked up at the airport by the GM of Smith Broadcasting's Ft. Walton radio station, WNUE. Holly Rogers was the trade-out king. Dean and I, being relatively new to broadcasting sales, were amazed when we stopped at a local beverage store and could fill a grocery cart full of various "soft drinks" and "put it on the trade." As we were turning into the motel, Dean noticed an excursion helicopter parked across the road and commented how he sure would like to fly in it. "No problem," says Holly. "Just tell the pilot to put it on the trade." Upon return from a 30-minute buzz of the beach, the pilot asked Dean for his $25 fee (things were cheaper than at the beach). Dean, now catching on to this trade business smartly replied, "put in on the WNUE trade." "What trade?" replied the pilot. Holly was also big on practical jokes.

Cactus happened to be down at the same time working on the radio station. While there he drove the station Newshound. This was a very old, very gaudy, VW Beetle. The motel where we stayed had a large sand strip separating the room wings. I asked Cactus for a ride around to the other wing. "Why go around?" he replied. We had to crawl out of the windows since the doors were sunk so low in the sand. While dusting the sand off his kakis, Cactus said, " Don't worry, we'll put the wrecker on the trade."

In order to stimulate more memory article submissions, I think I'll suggest to MD to give a very nice prize for the one he likes the best. After all, I can put it on the HiWAAY trade.

Newscasts

Three in particular come to memory from my 31 years at WAAY: Early Days…Johnny Evans was doing sports and came on in his brand-new Afro hairdo. I don't remember anything about his delivery or sportscast just "the do." Next MD did the weather using a mark on map. I knew either the fumes from the marker cleaner or his pre-newscast relaxer had gotten to him when he drew a tornado symbol for our area that wound up covering the better part of Alabama, Tennessee, Mississippi and Georgia. I'm not sure we were even under a watch at the time. As for the anchor, I don't remember his last name but his first name had to have been Mo.

Happy Talk…We had just aired a story about a couple winning a brand-new Cadillac in some contest and had switched back live for happy talk time before the break. One talking head commented to the other how lucky the winners of the car were. The other talking head agreed and said the happy couple told her they were going to trade the Cadillac for the Lincoln Continental they had always wanted. Guess the first spot in the break, yep, you guessed it, Sterling Cadillac. The Sterling owner called me at home before the spot was over to make that the last Sterling spot we ever ran.

Remote Broadcast...In my 31st year at WAAY, sitting in the remote truck at Guntersville's City Harbor, I gained a new sense of appreciation and pride for the young people around me. During a "WAAY Down The River" remote, everything that could go wrong looked to me like it did. Mics went out, lights went out, shots had to change, tapes didn't roll, you name it, it happened. The folks directing and producing didn't break a sweat. Talent on camera moved with the flow. Myself, I was a bundle of sweating tension. It took a couple beers and a hardboiled egg at the First Chance in Madison County to settle me down. I now understood why no one in the van was over 30. It was a great feeling for this old salesman to experience the admiration I had for the folks making it happen and the pride that I felt to be their GM for that last year before the new owners.

W2L

There have been some comments submitted that the W2L crew felt misunderstood. As happens often in broadcasting, that's a matter of perception. Engineering was actually the one misunderstood. We have to be reminded here that W2L was taking these engineers from Scottsboro a little too fast. As for sales, we were just jealous you were having so much fun without drugs.

With regard to selling advertising in W2L, the show was certainly prominent enough and was not too late. Some may have felt it was not late enough. Sales at first had planned to sell complete sponsorships. After viewing the first few shows, we changed the strategy to selling participating spots. Following the next few shows, this was revised to a ROS spot carrier. As creative juices really started to flow, W2L seemed like a perfect place to fulfill our PSA obligation to the FCC. After talking to several of the Public Service groups willing to give up their time in the show, Sales agreed that if you guys could give up so much of your free time the least we could do was give up the breaks and make W2L commercial free for our viewers pleasure.

Late Programming Note

For those of you who don't remember Diane Buffington, she was station artist for a short time and a good friend. One day soon after she and her husband moved to Atlanta, I received a phone call from her about a new daytime syndicated show that had just started in Atlanta. She thought we might want to take a take a look for our market. We passed. How could a show done in Chicago by a young female Afro-American make it? Especially, with a name like OPRAH.

October 1968 to October 1999. It was a great 31-year ride! How the time flew because we WERE having fun.

Joe Lowe, Engineering & Birth of HiWAAY.net

THE BIRTH OF HIWAAY

In the early 90's I had been using a local Huntsville BBS (Bulletin Board System) system called InterQuest which had internet e-mail, no web or anything else, just e-mail. Their connection to the internet was a single 64K leased line to UUNET in Atlanta. In the early 90's this was the only "Internet" available to the public at a reasonable cost.

MD had asked me to look into the possibility of setting up a WAAY-TV BBS so computer users could send messages to the station, download weather software, and in general be a place to connect with computer users which there was no shortage of in the Huntsville area. I thought that setting up our own BBS would be a real hassle, so I told him about InterQuest, and we contacted them about providing BBS service on a contract basis.

About this time, InterQuest had expanded their internet bandwidth and was offering full internet access. I had been playing around with a new program called Netscape which I had read about in Wired magazine. I told Mark Derrick about this new thing called the World Wide Web and Netscape and how he should come have a look. Being his usual self, he scoffed "I DON"T HAVE TIME FOR THAT!" I remember thinking maybe today you don't but some day real soon you will. I had better luck with MD though. I clicked on a few webpages and MD just looked at the screen and asked, "You mean you can get sound and pictures on this thing?" Yes, I answered, that and a lot more. He turned and walked out with not a word more.

Next thing I knew Mark Derrick and I were off to InterQuest for a little industrial espionage, orders were placed for equipment, and HIWAAY was born. WAAY with MD at the helm was always an innovator. We were the first to have stereo sound, Doppler RADAR, and many other things. The internet service was no exception, it was an instant hit.

Thanks, MD, for a great working environment where innovation was allowed to flourish!

The 31 Target B&W Station ID, then Color ID

I have added a color "target 31" ID from the 60's at the bottom of the main web page at www.31alumni.com. It was originally in B&W on an 11x14 inch art card, hand painted by Howard Troutman in shades of gray in 1963.

We had a "Coloring Contest" when we bought the used color film chain in 1965, and printed a "coloring book outlined version" in the newspaper and asked viewers to color it and send it in. The winner got a color TV set and we would use the colorized version of the logo on the air, which we did.

You are seeing a B&W version of that color ID, but we tested it with a B&W studio camera to make sure the contrast with the color looked almost exactly the same as the original version did for the many B&W TV sets that were still out there. For an example, medium shades of Red and Blue look exactly alike on a Gray-Scale or shown in B&W. That would have been a "no-no" in those days.

So, that's the history of the color target ID. My father liked the term "First In Huntsville" as 31 was the first station on the air, but it gave the impression we were #1 in the ratings, which was not always the case.

My thanks to Val Ginter for sending it to me. (MDS IV)

Adrian Gibson – News then Weather, 30 Year Span

So many things have happened, evolved, crashed, that it's hard to choose which topic to submit.

Let me go back to 1970. I had been at WAAY radio almost three years when MD Smith III called one morning and insisted that I fly to Ft. Walton, Florida with him. I pleaded a busy schedule but Mr. Smith would not listen to that as a valid reason.

Anyway, as we were enroute to the airport, Mr. Smith asked me, "How's your stamina?" He said he wanted me to go to TV but still do morning news on radio. I honestly did not wish to go to TV. Never gave it a thought. Mr. Smith said if I would go to TV news and didn't like it, or if he didn't like the things I did there, I could return to Radio.

My first newscast on 31 was December 26, 1970. And, yes, it did take a lot of stamina to keep afloat. It has been interesting. It would take a complete book to talk about the unusual people, places and events over these 32+ years. Spent the last years doing weather on Channel 31.

I look forward to seeing the "old crew" next month.

Adrian

Helen Howard Carroll – Reporter-Anchor 1980-85

Surprise party, Dick Van Valkenburg and other events

One of my favorites was the night we gave REX (and what WAS his last name) the live truck operator a surprise party.

Dick Van Valkenburg made elaborate arrangements with fire, ambulance, and police to set up a crime scene, complete with body at an apartment complex (where a 31-staffer happened to live). We called REX out to the scene...and when he got there, set up the truck, grabbed a camera, sirens were wailing, lights flashing. Suddenly, the body jumped up to greet him (anyone remember who played the body?) and we all jumped out of bushes and from behind trees. It was great!

Speaking of Dick—there was the terrible day when he had his first heart attack in the newsroom. About 4 p.m., just after the last call for the film run on the news. He was in the audio booth recording his track but had the presence of mind to open the door before falling to the floor. Wayne Matthews, I believe, performed CPR while waiting for the ambulance. I was out on a late story...and arrived back in the newsroom as they were carrying the gurney down the stairs.

I'll send more as they bubble up to the surface. Again, I'm so sorry I won't be there...please keep me in the loop and on the mailing list with all the pictures, etc. I would LOVE, LOVE to have a copy of the memory tape you play. If I send you an envelope and tape could you make me a dub?

Lu Ann Cahn – Reporter, Anchor & Olympic Hockey Score

What a blast from the past! I'm so thrilled to hear from you.

I must say I think it's wild that Rick Davis and I share the same most memorable event at WAAY! The night I announced the score of that Olympic hockey game I thought would be the end of my career. People in Huntsville were so furious with me for revealing the score. I was so green at what I was doing...such a rookie. I remember I made an apology on the air that night and was happy you didn't fire me!

I have another memory of coming to your office to tell you why we had to get rid of film and get into video. You were heading there already, but I was venting about how I ran out of film during some important story and missed the critical part of it. Do you remember any of this? You patiently listened to me.

Do you know all these years later, I treasure all those experiences being part of film during its last days in news...working with the very first live truck at your station...anchoring with Dave Stanley without a teleprompter...no computers...I was so young and impatient which is why I probably left too soon...I made so many mistakes and would make so many more down the road before I became a so called "polished professional."

I want to thank you for that year...1980...thank you for taking a chance on me. I'm truly grateful and wish you and your family all the best.

Sincerely, Lu Ann Cahn

(Later email update)

You know what M.D.? I'm very lucky. I've been married 21 years (to a TV news photographer...) We have a 16-year-old daughter and I've survived cancer and we've made a nice life for ourselves...I feel very sentimental about my past and all the stops along the way (or in your case, WAAY.) You've helped a lot of people over the years realize their dreams working at your station...Enjoy and celebrate!

Lu Ann

Sam Arnold – Many Jobs, 1959 BC (Before Cactus)

[Editor's note: Sam was/is a multi-talented guy. He directed news casts, Romper Room and even taped shows like the Schlitz's (Joe Adams) "Gone Fishin'" show. But Sam also shot film, slides, processed film and slides (all B&W of course), mounted slides in glass for air use, shot polaroid slides for graphic use on the 4x5" rear screen projector slides (that projected this image on a screen behind the anchor), wrote and produced commercials, did switching (which we call master control work) and about anything short of heavy engineering. He had a mild temper in those days but was still careful and accurate in his work. He was one of the ones we REALLY hated to lose in the 60's when Sam felt it was time to move to a better job.]

Thanks for contacting me. I look forward to seeing many of the old group. I am BC (before Cactus) and AJ (after Johnny).

Here are two of my war stories.

Time was fall 1960 with station operating on 1000-watt transmitter and 150' tower. It was about 10:45 p.m. and raining like crazy. The phone rings and a guy asked if we are on the air. I asked where he was calling from? He said Madison. We were running an awful late movie, (understand this is before Smith Broadcasting) so I suggested he go to bed and forget it. I said our signal was falling off the side of the mountain so if he did not live at the foot of the mountain ain't no way he could see it.

Oh well, one more quick one.

I grew up in Memphis area and attended Bartlett high school. A school buddy moved to Huntsville shortly after I did. Johnny or others would tape remaining station breaks and introductions after the 6 p.m. news. I called my buddy and asked him to watch around 10:30 p.m. As Johnny introduced the movie, I gave him my school class member's names as stars of the late movie. *(So, class member's names were listed as all the stars of the movie instead of the actual movie star names.)* I attended a class reunion many years later and my friend reminded me what happened. I think Johnny and I were the only ones who knew what happened.

Thanks, Sam

[Images before Smith Broadcasting bought WAFG-TV in 1963. These are earlier days.]

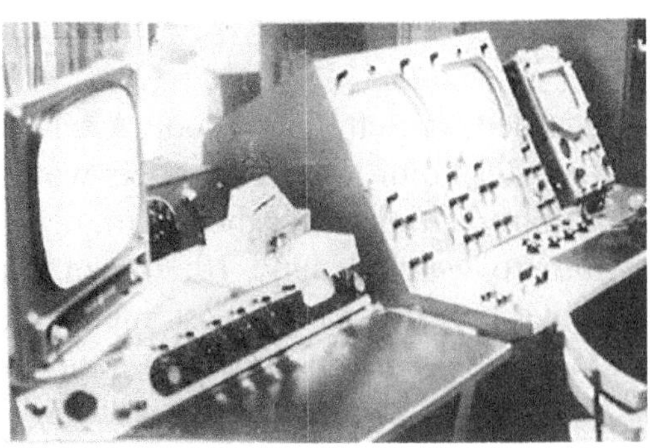

1981 Ratings, Anchors & "Bucket 'O Bolts"

31 EYEWITNESS NEWS

The Winner Again
In The New Ratings!

OF THE 18 - 49 YEAR OLD ADULTS WATCHING THE THREE HUNTSVILLE TV STATIONS, THE SHARES ARE:

6-6:30 PM NEWS	
WAAY-TV	50%
WHNT-TV	29%
WAFF-TV	21%
10-10:30 PM NEWS	
WAAY-TV	45%
WHNT-TV	28%
WAFF-TV	27%

At 31 Eyewitness News...
The Real Winner Is You!

SOURCE JULY '81 NIELSEN RATINGS SURVEY

This is the news team of (L to R) Jamie Cooper-Country Rover, Bob Baron-Weather, Kelly Cooper-News, Jim Marsh-News, and Dave Cody-Sports.

These 1981 Ratings were just terrific. There were a number of years that we beat Channel 19 by a wide margin (before Channel 48 became a contender). This anchor team is representative of the strong and talented people we assembled over the years. There were many changes and sometimes it took a while to regain our #1 position again, but through research and tireless efforts to be Number One, it happened. It took the entire staff, not just the anchors, however. Promotion had to do their job advertising and promoting these people and the newscasts. Engineering had to keep equipment working, both on-air as well as the cameras and recorders and live equipment. Production had to produce slick, error free newscasts that showed the audience why we were the leading station. We had to resist cramming in too many commercials when we were #1, so as not to run viewers away and leave room for content, but it was nice to make the money while we could. It helped us continue to grow and give raises. Winning is always a total station effort taking every single person.

Channel 48 and Channel 19 had moved down the mountain, while we continued to expand and/or patch up the old, original building on the mountain. We were referred to as "That Bucket of Bolts On The Hill" in one competitor's meeting. They were saying, "How could 'That Bucket of Bolts On The Hill' beat a station with state-of-the-art facilities, building and location?" The answer was, it's what the audience sees on the air that matters and the morale of the people at the TV station. We spent money for on-air and for morale because that's what mattered, not how pretty and new the building was.

I can remember many a department head meeting where arguments would break out about conflicts with news, engineering, production, traffic, promotion and every other department (including janitorial). But it was because we all cared and wanted everything we supervised to be the best. That's what it takes to win. It takes caring, a strong desire to win and a lot of heart (Ask Sea Biscuit).

M.D. Smith, IV

John Malone – Radio & later TV work

A brief thought for your "memories" page.

2003 marks my 40th year on Earth and my 25th year as a broadcaster. I grew up in Fayetteville, Tennessee and cannot remember a time that WAAY Television was not a part of my life. Even as a preschooler, I was fascinated by radio and television. One of my earliest memories was visiting Channel 31 with my father and brother and appearing (briefly) on "Coffee Break" with Johnny Evans and Mary Beaton. My dad was acquainted with Johnny and was able to arrange a studio tour. We were allowed to actually sit on the "31 News" set while the program aired live–what a thrill! I remember touring the facility and meeting Adrian Gibson. He told us something like, "This is a newsroom. You're really not seeing us...we're ghosts!" I never understood what Adrian meant until years later when I got into the biz...news people are supposed to be out in the field covering stories, not hanging out at the station.

Several years later it was a thrill to be a part of the WAAY-TV/Radio team (in radio production and, later, TV weather). There is no doubt that WAAY served as a significant and positive influence in a broadcasting career that has been very rewarding.

John Malone

John Keel – 1978-1985 – Film and What Didn't He Do?

John Keel (1978-1985)

Gee, where do I even start? WAAY was my very first official job and I was there for seven years (with one short break - more on that later). There are SO many memories - almost all good - that I may actually have forgotten more than I can remember, especially now that it's some…what? Sixteen years since I left? Wow. Of course, then there are those stories that I do remember, but will carry to the grave. (And you all know who you are!)

The first thing I remember is actually getting the job. I was going to Calhoun and had classes with Dave Rowley and Ken (can't remember his last name now–red hair, moustache, news photog–was it Hoyle?) and they got me an interview with Rip, the Production Manager at that time (NOTE–from now on, if there's no last name it's because I just don't remember it–sorry everyone!) I started part-time running studio camera and was SO nervous when I finished training and had to do it all on my own! I also remember thinking I was super cool because I worked at 31.

Another thing I remember is what Liz Webb Cernadas (hey, Liz!) mentioned about having to hand draw Bob Baron's weather maps! It seems so incredibly antiquated now. Plus, there were the occasional temper tantrums and fist fights between anchor people! Ah, Show Biz.

From weekends, I moved to weekdays and also began working as part-time projectionist (oh yeah, we shot FILM in those days, kids). Nothing will stress you out more than having a photographer fling you a reel of film that's late and the anchor is reading the intro and you've got to thread that RCA projector in about 5 seconds!

From there, I started working in the film room, splicing together movies and cartoons and then started working in production during the day in the booth. I don't remember how I started, but I managed to do it all, mostly under the watchful eye of John Stanners–audio, vidifont (oh God, remember those big 6-inch-ish discs with like, maybe four fonts per disc?), and ultimately directing. There were days I remember when I would direct Morning Waay completely by myself! I would roll my chair across the floor to get from the audio board to the vidifont-hitting e-mem buttons on the Grass Valley to fire off fades or effects …almost never missing a beat.

There were also commercial and promo shoots, working with Carl Spurlock and Jeff Davis on location and in the studio. Looking back, there was rarely a dull moment. Plus, at some point there were stints on master control, mostly during the afternoon soap block, which wasn't too stressful, but addicting. (Can you believe Luke & Laura are STILL on "General Hospital?" Not that I watch or anything.)

About five years after I started at 31, a big-time Hollywood movie (or so we thought) came to town and I wound up getting a job on the crew in the camera department. I thought this was the big time, baby, and so I left 31 to work on "Secrets of the Phantom Caverns." I don't know why I thought that Hollywood was a well-organized, perfectly oiled machine, but it by no means was (and after living in L.A. for 7 years now, trust me–it still ain't). It was a mess of a movie and they almost killed all of us by filling up the cave we were shooting in (wasn't it Cathedral Caverns, east of Huntsville?) with carbon monoxide by running generators for the lights. So, I left the movie totally disillusioned with movies and making it in Hollywood. A few weeks went by before I managed to get a job back at 31, this time in the news department as a photographer.

That led to quite a number of adventures as well, but I've gone on way too long as it is, so I'll save those memories for a sequel.

Of course, what made it work during all the years was the people (some of whom I've mentioned already), some of whom I still see or keep in touch with, but all of whom I remember as being helpful, encouraging or just plain fun in various combinations and degrees: M.D., Bettie Higgins, Lew Koch, John Stanners, Liz Webb, Penny Norwood, Deb

Napolitano (who is living in NYC), Donna Spurrier, Jan Miller, Lee Baker, Ron Milner, Bruce Hutson (who took my recent wedding photos, see below), Bill Bocking (he's in Louisville), Liz Hurley, Mark Smith, Alan Dunkin, Debbie Benson, Lou Ikard, Missy Garrett, Eric Eisgrau, Brenda Wood, Rick Davis, Adrian Gibson, Jamie Cooper, Jeff Wigginton, Jerry Carnes, Joanne Powell, Ron Brooks, Bob Labbe,… I'm sure I'm forgetting so many people, so I apologize to anyone I didn't mention.

I'll close with some quick bits as I brainstorm through random memories:

> The Santa Show! With the exception of whatever leggy Miss Merry Christmas we had that year, it was always a pain in the ass and I hated Christmas for years because of it! Oddly enough, I have no memory whatsoever of who played Santa.

> Adrian calling me to wake me up when I was directing the early morning news and would oversleep. Luckily, I only lived about ten minutes from the station then!

> Two different Ku Klux Klan memories: the first was driving up to work in the fog one afternoon and suddenly seeing the robes of KKK members protesting the station when we showed "Roots II" appear out of the fog—THAT was pretty eerie. The second was when I went to help the news team cover a huge KKK protest/rally in Decatur and Eric Eisgrau picked a spot BETWEEN the National Guard and the KKK to stand and get some shots. That was nerve-wracking, to say the least!

> Leaving the "ON AIR" light lit up after we went off the air on the days we had cooking segments on "Morning Waay" so the crew could get to the food before the rest of the staff!

> Rex Free's big surprise that Helen Howard mentioned in her piece. That was me that played the dead body. I remember Carl Spurlock arranging my limbs on while I laid on the concrete so I would look suitably dead and then they poured fake blood in my hair and down the sidewalk. One of the funniest pranks ever. I saw Rex not too long ago and we were talking about it. Still a great guy, that Rex.

Okay, that's MORE than enough. I've thrown in a few photos here at the end and actually have more somewhere that I'll have to find and scan. I hope that the reunion goes well and I can't wait to see photos from the day.

All the best,

John Keel (does anyone even remember me?????)

Lisa Greer (Ceci) – News, Anchor & "Wall to Wall" Coverage

"WAAY To Cook"
"Finger" Food

WAAY to Cook was the favorite feature for all of us involved with the midday news, because after the show we got to sample the treats! During one of our cooking segments, the young chef from a local market quickly and nervously sliced vegetables with a long knife and tossed them into the dish he was preparing. My co-host (I believe it was Al Whitaker) and I were alarmed to see blood squirting onto the cutting board but continued on with the segment. When we went to commercial break, the chef used a towel to wipe the blood from his cut finger and we could all see that the tip of his finger was GONE. We all looked disappointedly at the dish that could have been lunch.

Word travels fast and the crew that usually swarmed the "WAAY To Cook" set after the newscast didn't show up to sample this finger licking treat.

WALL TO WALL

My greatest memories at WAAY TV include breaking news events, especially "wall to wall" weather coverage. What a great feeling to be the first station to bring people the news they want and need to know, and to do it better than anyone else in the market!

My first "wall to wall" experience was during the unlikely snowstorm of March '93. We went live from 6-10 a.m. I learned that morning to go easy on the coffee while sitting at the anchor desk, because it might be hours before the next break.

Other weather memories include: Kevin Athow and I surviving as a funnel cloud passed over University and the Parkway, (I thought I was surely going to die right there under that overpass, and I learned that when Kevin says, “Hey, there’s a funnel cloud,” he knows what he’s talking about), an unnamed photographer putting up the mast as lightning moved closer so we could “go live,” and meeting at Exxon for the cold, crowded ride up the mountain in the vehicle with chains.

I’m thankful for lasting friendships formed during my time at WAAY.

These days, I’m busy taking care of triplets. My three boys are two years old and the new loves of my life.

Bruce Hutson – Photog in Wind, Rain, Snow & Live Ammo

Today and 1983 Winds of War party

Well, let's see how a story or two might unfold...

I came to Channel 31 in 1980, fresh from Miami, Fl. I was kind of passing through town, visiting with my Mom for a while, saw an ad in the Times for a news photographer (no shooting experience needed, still photography a plus) and came up the mountain to talk with Mike Sullivan, then the News Director. I had only my still photography portfolio to show my "eye," but Mike went ahead and hired me. We were still shooting film with CP-16's (talk about way back machines), and just transitioning into video with the huge TK-76 camera and a "portable" 3/4-inch deck. Try running up and down the sidelines of a Friday night high school football game with 40 lbs. of gear strapped around your body! Ah, youth! I was kicked up to Chief Photographer, produced the weekly medical feature "Eye on Health," with Dr. Charles Morley, and also shot the still ads for TV Guide, and the infamous Hall of Shame Anchor portraits. As M.D. said when I asked him why I had to shoot stills when I was swamped with duties in News, "we do whatever it takes." And we all did. As I look back, I can't believe all that we accomplished in the short two years I was with the family. From unbelievable lives shots at the Space and Rocket Center, racing downtown for a possible jumper from a water tank with Adrian Gipson—Mike Sullivan later told me he wondered if

I would follow the guy if he jumped. Luckily, I didn't have to find out, though I know the answer. Snow storms and camping out at a reporter's apartment so our team could cover whatever transpired, working Thanksgiving or Christmas (new doesn't take a holiday, whoever said it), running as part of the 31-relay team in the marathon, doing commercials and promos with Carl Spurlock. (Hey Carl, remember the bathing suit one and the make out silhouettes for Gorilla theater?) Adrian's best line on camera, "if and buts were candy and nuts, every day would be Christmas." I don't remember the story but it came out of nowhere and I just lost it.

Filming Helen Keller's home, Opryland and Graceland with Steve Newvine for One Day Getaways, a personal favorite because we were the first media allowed into Graceland before it was opened to the public. This was 1981, just three years after the death of Elvis. Steve, a HUGE fan, suddenly channeled the King when we were left alone, unsupervised, upstairs in Elvis's bedroom, just next to the bathroom where he died. I sort of remember Steve opening a closet, trying on one of the King's capes, and cutting a move—Thank you, thank you very much.

Of course, Winds of War hoopla. Those pics of me are scary, thanks Helen.

Oh, yeah, the mercenary training series that garnered an Emmy Award with Jim Marsh reporting and Jim Miller producing. We spent several days with those wackos somewhere near Selma in the dusty oppressive heat of the summer. They trained their paid participants with live ammo, so of course since Jim Marsh was the on-camera guy, naturally he had to "imbed," and take part in all of it, and I mean all. We camped with them, ate their food, got bit by the same damn bugs. I rolled tape as Jim waded through water and crawled on his belly as LIVE rounds from an M-14 were fired in his direction. I paused after one particularly sweaty, hellish, moment for Jim and told him we had to go again because I ran out of tape in mid-hell (for him). Let's just say if looks could kill—no bullets in his gun—I fessed up that it was all good. Hey, whatever it takes. Did I mention LIVE AMMO!?! Man, what the hell were we thinking?

Eye on Health with Charles Morley was great. I scoured medical journals for material and one day each week I would disappear all day to shoot with Charles. While Mike or Jim Miller kept calling to see what was taking so long, they thought it should only take three or four hours, Charles took me to lunch, we hung out and talked sports, literature, film, travel, currant politics, then shot the piece—a very civilized way to produce. The weekly feature was top rated, and Charles loved the attention, though he wouldn't admit it, and all the docs in town readily opened their doors to us.

I was in the film room the night of Dick Van Valkenburg's heart attack that Helen talked about. "Souping" the film was a solitary task that I enjoyed—just the radio playing in the dark for company. When I opened the door that night, I saw the EMT's as they rushed by with Dick on a gurney. It was very surreal to step into such drama without any indication beforehand of what had unfolded the previous 20 minutes. I think that was the last hurrah for him at the station and he was relegated to rule his spot news roost from his recliner at home. He would page the photogs on call at any hour of the night to cover the news. I was awakened one late Saturday night (actually early Sunday a.m.), to drive to the River by Ditto Landing, body in a car trunk. Let me tell you, when a pager goes off at 3 a.m., serious thoughts of ignoring it drift through your head, as it did happen on more than one occasion. I called in and Dick even said, "Didn't think you would answer it." But I did the drive from north Huntsville, got the body out of the trunk shot, and also got some kudos from Dick, which was pretty damn cool to come from him.

OK, one more, the last 31 moment I have and it took place a couple of weeks after I quit work at 31. Bill Bocking, Homer Hickam, a scuba diving buddy and author of the book Rocket Boys that was made into a movie recently, were water skiing off Ditto Landing when The SCI-Tanic sank. This was a terrible tragedy where I think 18 or 20 people drowned when a huge stormed upended their boat—31 followed up the story several years ago so this all came back to me again. Bill, Homer and I were among the first to arrive in the middle of the river after the storm passed at the now upside-down boat. We all hoped for a rescue operation, but it soon became obvious that all aboard were lost and we spent the afternoon pulling bodies onto the flat underside of the boat. Homer dove repeatedly with borrowed scuba gear, located everybody inside, and gently floated them to the surface where Bill and I lifted them topside. It was traumatic for all involved and we knew that the circling boats contained news crews filming the activities. Bill and I talked about being on the other side of the story, how one's life can change in a split second and how we all changed that day. When we pulled up to Ditto Landing, we saw all the Fire Trucks, ambulances, and news vehicles and the last thing we wanted to do was face a camera. I ran into Steve Kerr, he caught my eye and knew something was up. He asked if I had been out there. I nodded and knew his next question. I said no way could we speak about what had happened, and to his credit (at least to me), he said OK. The next day I was up at the station picking up some of my remaining stuff and the newsroom was buzzing with the previous day's coverage. I still hadn't told anyone I was involved and the air of distance in the room from that moment was obvious. Talk was of selling the footage to network and CNN, sending out follow up crews, just the normal newsroom stuff. I finally pointed out that the figure in the green t-shirt in the footage was me. It was cathartic to talk about it with my former work mates, but I knew I was no longer part of the team and it was best to leave them to their jobs, which they did exceptionally well. 31 had the evenest coverage, and I watched it all from the comfort of my house. Homer still lives in Huntsville, and Bill is up in Louisville, a Fed EX pilot, good guys both.

Interesting people, stressful moments, a big-time learning environment at 31. Although I don't shoot news today, thank God, the work ethic of getting the story under deadline stays with me.

Kelly Cooper – Dead Body and DDT in Triana

1981 & 2003 photos

Typical news hound -- I wait until DEADLINE before filing my report!

I will always treasure my days in the Channel 31 newsroom. It was extremely intimidating following Brenda Backmon Wood in 1980, and I vividly remember my first night on the air. I was so nervous my voice cracked as I attempted to introduce myself, and all my dear colleagues from my previous station, channel 48, telephoned after the show to imitate my "squeak." We had a great laugh and, subsequently, a wonderful two-year run. My co-anchors during that time included Bob Baron, Jim Marsh, Rick Davis, and later Dave Cody. Mike Sullivan was an excellent news director and managed to cultivate a strong team of reporters with an incredible amount of energy. Wayne Matthews worked tirelessly (and often doggedly) on the never-ending controversies surrounding Police Chief Sal Vizinni. He also resuscitated Dick Van Valkenburg when Dick had a heart attack in the newsroom's voice booth. There was our capable producer, Jim Miller, who later went to WSB and CNN. Jim Keller, now communications director with Boeing, was the Shoals Bureau chief and later anchored weekends with Helen Howard after the delightful Steve Newvine departed. Jamie Cooper was the Country Rover, and naturally, everyone thought we were married or siblings or "kin." (FYI—we weren't related in any way.) I will never forget the sly and dependable Eric Eisgrau or the wild and wooly Ace Askew.

So many of my funniest memories, unfortunately, I can't repeat. But there were poignant moments, such as the time as Deputy Dawg, our infamous cameraman, and I returned late one night from a city hall live shot and found a young man's body in the middle of highway 431 near the overlook. I recall extensive work on a DDT–Triana series and another on TVA's nuclear power build-up, the latter prompting a succession of threatening letters from laid-off TVA workers. I recall falling shards of ice from the tower crashing through our car windshields during unusually cold winters and reporters scrambling to move cars when Baron called for more freezing moisture.

Mostly, I recall the wonderful camaraderie and teamwork during those two years, not only in the newsroom, but among the production crew, art department (Lee Baker and Bruce Houck), sales department, and engineers. The ratings were high, the employees professional, the nights long and filled with longer tales that emanated from bars that stayed open past 2 a.m. (remember the Hilton?) It was a great time in our lives—Many thanks to MD for organizing this reunion!

Kelly Cooper Schrimsher

Dave Stanley – 70s Serious Joke with Truth in It

Little known story - until now.

[Editor's note: This was written after the reunion. I don't clearly remember it, so it must not have made too much of an impression, but I do remember anchors doing some weekend duties, so I don't doubt the story.]

Ken Rainey was a little upset because I refused to share this story at the reunion. It is something that happened in the 70's known to only a few. It involved, in addition to myself, Ken Rainey and Dan Jamison. Tony Beason may have been involved, but I can't remember. Adrian Gibson was not involved because, at the time, he was the boss. I may also add a damn good boss too! Hey, this guy not only knew where all the dirt was, he knew where all the dust was. Talk about an amazing resource.

Anyway, it is a tale I hesitated to tell because of a real concern about Mr. Smith upchucking that delicious BBQ. Plus, given my fondness for Mr. Smith and his family, I still feel an element of guilt. So, for the first time ever. Let the truth be told.

Most won't remember this, but waay back when anchors had to work on Saturday. So, sitting around one day in the newsroom we decided enough is enough. Not to brag, but back then we were all household names with incredible ratings and carried a bit of weight. Albeit, most of it in our heads.

We decided to go into Mr. Smith's office and demand to be taken off weekends or we were all going to quit! His reaction, with his face turning a scary pale, was immediate. Within two weeks we were all off weekends.

Now, the real story.

Incredible as it may seem, given Mr. Smith always being on top of and into everything, he neglected to do one little thing that day—look at the calendar. You see, it was April the 1st. April Fool's Day. The whole thing was a JOKE!

But seeing the effect of this dubious, rather cruel and childish attempt at humor as we glanced at each other the same thought passed among us. What the hell, now is NOT the time to back down.

Should we have been forced to be the only anchor team in the city to work on Saturday? Probably not. Should we have accomplished what we did the waay we did it? Okay, we were jerks. But then it was a joke. Or, at least, it started out that waay.

Whew, now that that's off my shoulders I feel a little bit better about myself. Truth be told, however, we also felt pretty damn good at the time about not having to work weekends any longer either.

Mr. Smith, I truly felt and continue to feel sorry about causing you to almost go into cardiac arrest. But it was, as God is my witness, a joke. Not funny, but there it is… Another slice of WAAY-TV folklore.

With much love from Irene and me to you and your family and all of the WAAY-TV family,

Dave Stanley

[Editor's footnote: I do know that I learned to always take people seriously. A few times in my younger years, I might laugh, assuming someone was joking when they were not, and

that had bad outcomes. So, even if I might have felt they were joking, I would have handled my response with complete seriousness. And as you see, even as an April Fool's joke, there was seriousness behind it. I also had been approached times before and after these years, where I was put into a bind and a key person would threaten to quit unless his/her demands were met. Usually, they got their demands met until I found replacement personnel. In this group's case, they were not serious about quitting, but did want off weekends and we did find ways to work that out. MDS]

Hal Comello and WAAY Radio of the 60's

In this photo taken in 1961, you see (left to right) Henry Beam (Manager), Dick Essner (Engineer/DJ) and Hal Comello (Program Director and DJ). The woman in black is Henry Beam's wife, Peggy. The man next to her is Dan Akens (News Director and DJ) and two of the salesmen.

Hal Comello did a really great morning show each day and he was a lot of fun to listen to. He was always bright and cheery, and he was also that way OFF the air. He was often caught in the middle as my father, M.D. Smith, III would call and "chew on him" about one of the young DJ's playing the music out of order or playing their own records that they liked. I can still remember Hal musing something like, "Oh, well, M.D. is really chewing on me today about (name of disk jockey) and I guess I have to do something."

The photo at the top of the page as well as the larger version below, is Hal these days as he works as a TV news anchor in Mississippi.

Of the group photo and working for my father, Hal said, "The year was 1961. Can you believe it, 43 years ago? I will think about some of the memories of my days with Smith Broadcasting. I remember a lot of things about your dad who really treated me great. One was when I was trying to buy a home in Huntsville but needed $300.00 for a down payment. He loaned it to me without any hesitation. Also, one time your mom I think was doing payroll and somehow, I received two weeks salary instead of one, M.D., and your mom both said not to worry and to keep it as it would be too much trouble to change the paper work and make out another check."

M.D. Smith, IV – '63 Purchase Deal, Kennedy, and Managing

Memories of special times at WAAY-31-TV from 1963 to 1999. From Operations Manager to General Manager & CEO.

Here's a sample of business cards I had from 1962 at WABP student Radio Station in Tuscaloosa, where I was a salesman one semester.

Also, the first business cards I had at Channel 31 in 1963 (top two) along with a host that followed that over the years. Logos changed, card color and style while always keeping a basic red, white & blue color scheme after that very first temporary card with the TV camera on it. The first real card we had, the big 31 with the call letter through it was hand painted by Howard Troutman on a large art board, and we photographed it, shrunk it down, and used it both on-air for station IDs as well as our cards.

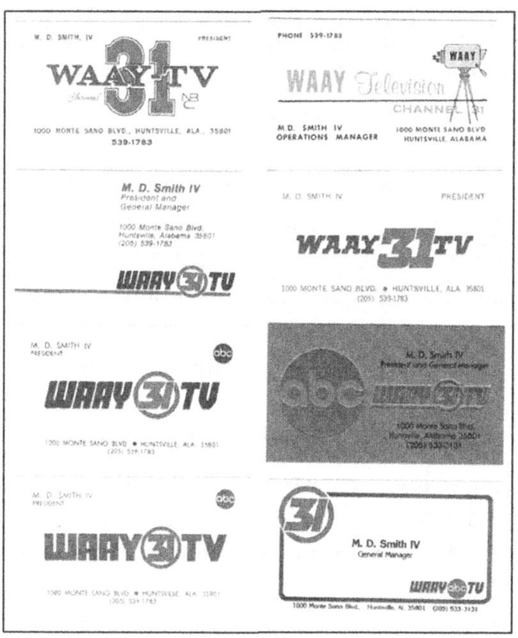

1963: We had planned to put Channel 25 on the air from scratch on the parcel of property we owned that later became Ch. 19's transmitter building. But the owners of Ch. 31 seeing two new stations being built at the same time (19 and 25) decided it was time to sell, so they negotiated a deal for $500,000.00 to purchase it and 20 acres of property.

I remember my father and I dealing with Big Jim Beasley, owner of Sweet Sue Kitchens that canned chickens and dumplings in Athens. It was a big operation and Ch. 31 was just a hobby for him. He was the majority stockholder and while some of the local management and minority owners in Huntsville did not want to sell, when Jim told his buddies, "we're gonna sell that station to M.D. Smith," it was a done deal. We turned over the check at his office in Athens in July 1963. He was late because he was tending to a sick race horse, worth $20,000 and that was a very lot for a horse in those days. He showed up in old jeans, boots and work clothes and certainly didn't look like the millionaire I had expected to see.

It was going to take the FCC about 3-4 months to approve the deal, so we just had to sit tight until we got the approval. In the meantime, I continued as Merchandising Director at WAAY Radio, did some late-night DJ shifts because I had a First Phone FCC license required for directional arrays at night and met one of the engineers at Ch. 31 named Bob Gay. I got my ham license at Tuscaloosa in 1961 (WA4DXP) and was glad to find out that Mr. Gay (Cactus) had one also. We had a lot in common and told lots of stories about building our own equipment. Mr. Milwee was Chief Engineer at Ch. 31 at the time.

The FCC approved the sale in early November of 1963 and we took over that operation of 17 people. Everyone did multiple duties. As Operations Manager, I wrote copy for advertisements and promotions, learned to sign on the station and get all the tube equipment up and running for the day (much of it would NOT come up and had to do troubleshooting with a tube tester, or direct substitute with another same type tube until the trouble cleared, then throw the bad tube away and return all the other new tubes to their cartons.) I managed all departments except sales, and Maury Farrell was hired from Birmingham to fill that role. At that time, I believe the total year gross income from sales was about $240,000.00 and the station was not operating at a profit, but nearly break even.

Then, later in November 1963, President John Kennedy was shot. All stations started carrying network coverage (we had ABC the 3rd network at the time) and NO commercials for FIVE days. That really hurt a new station just in the peak of the Christmas buying season so that November was the worst ever in the history of the station.

I had graduated with a BA degree in Radio and TV with a minor in Commerce from the University of Alabama in May of 1963. In college I worked for WNPT and WJRD as a first-phone engineer. I also worked for the student AM station WABP as the Sales Manager.

(This was just before Gov. George Wallace "stood in the school house door" at Grayson Hall at the University of Alabama in June and it was broadcast nationally.) I had spent a number of years directing shows, writing shows, running camera, control boards, audio consoles, printing super cards on the type setting machine and learning FCC rules and regulations from other classes. I knew all the technical things I needed to make a small station run and did many of these jobs (including talent of a number of occasions) as the need arose. Either we had a turnover, or someone was out doing something else, like shooting slides on location for a commercial, so I had to TD the newscast for Sam Arnold when he was not there. Sam was a top man on processing B&W slides and changing chemicals in the B&W processor and running the raw film through it when the news guys (2 of them) returned in the afternoon with their stories on film. We had one OLD 16mm Auricon Sound-On-Film camera that even used battery powered tubes in the associated audio system and motor drive. It was a monster and it cost a lot to buy all those different "A" and "B" batteries. The other camera was a silent 16mm Bolex with a turret of 3 lenses on it for wide, normal, and telephoto work.

The director's booth had a very simple switcher with two banks of eight inputs and a set of fader bars. That's it. You could put one camera on a black art card with silver lettering and by pulling the fader bars in opposite directions get both cameras on the air and you'd see the white-looking titles "supered" (superimposed) over the other camera, slide, film or whatever source you had selected. There were no "keys" in 1963 for us. A few years later we did buy an OLD, used RCA tube "keyer" that allowed for video keying and some special effects like split screen, boxes, etc. It stood over 3 feet high in an open rack, solid tubes top to bottom, and was as cranky and temperamental as hell. Cactus could only keep it going sometimes, and it ran up the monthly tube cost budget, which was one of the biggest expenses we had in those days. The cost of film and chemicals was the other large costs. Some years later when we went color film and slides, it really got expensive.

My Managing Experience(s) The one type of training you didn't get in school was management (no, not even in marketing and commerce). That is the process of controlling people, getting desired results and trying to keep everyone happy. My father never had schooling on this topic, so a lot of my earlier methods I got from him. He was the one that told me I didn't have to be liked to be a manager. In fact, he said, it would probably be good to be disliked, then you get things done when you say to do them—or else!

I was not inclined to feel deep down, this was how to manage. After all, I had worked for several radio stations in Tuscaloosa while I was in school (WJRD and WNPT) and I knew nice managers from horses' asses and I felt like I not only enjoyed pleasant managers, but that I would produce more results for those I liked, instead of simply doing what I was told and let the cards fall where they may if something screwed up.

But my dad was the owner and my boss, so I did my best to do what he said. We did keep expenses down, and Cactus (bless his heart (as he would say)) was one of the world's best at making do on a shoestring and used or free equipment. He still is in 2003. So, we kept expenses down, sales were not bad and we had a bit of prosperity. It seemed like a lot of it kept getting used up when we had to have more than one video tape machine. We bought the station equipped with one Ampex VR-1000 monochrome 2" video tape machine. It was a monster with pumps, motors, servos and hundreds of tubes to make it work. In addition to what was in the housing of the machine (4 feet by 3 feet by 6 feet) there were two 7-or 8-foot racks full of tubes, transformers and power supplies for that beast. There was no electronic editing, but we didn't have a second machine to edit from anyway. So, one of the biggest expenses we had was buying two used and reconditioned 2" tape machines. The TR-70 was built for color and was a "Cadillac" by our standards, and in addition, all we could afford for another, was a used RCA-TR-22 that had been "colorized." More about going color later.

Back to "managing" the staff. I tried being "not-liked" as a management style and things were sure quiet in the station for several months. A couple of the employees, like Carolyn Smith and Cactus said they liked things better when I was not trying so hard to be not liked. I felt the same as they did, and gradually let down and went back to being one of the boys and girls in the station, and at the same time, having a bit more fun in the TV business. I enjoyed learning and doing every single job in the station, but sales. I was not cut out for that, and I was glad Maury and his several salesmen, took care of that end. If sales were down, HE got chewed out for that, not me.

So, I didn't really start to manage anyone until late 1963. In 1964 and part of 1965, I realized how much I didn't know about managing people and how different they were. What motivated one and got them to work with a good attitude, didn't work on someone else. I needed to know more about people. About this time, John Patton of Dale Carnegie Courses came around. He had talked to me once that summer at WAAY Radio, but my father would not even discuss the fee for me taking the course. John came by to pitch it to me again, and this time I was ready, but again, my father turned a deaf ear to my recommendation for tuition. John did tell me a lot of the course was based on "How to Win Friends and Influence People," so while I could not take the course, I could buy the book and read it. That changed my life. I was so enthusiastic and I tried parts of the book and they worked. Now I had to have the course, and we finally worked out a TRADE where John got advertising for the program and I got to take the course. Another revelation over that 14 weeks in 1965. I learned a lot about how people "ticked" and what things motivated them and what things turned them off. You will have to ask people who were with me during those periods about the changes. Cactus remembers. It made my job so much easier, even made it fun. I saw people's attitudes change and real teamwork begin to grow. I also recognized when some "prima-donna" was hurting morale and I worked on changing those things or "let them find someplace else to work."

The advent of color. WE were the first TV station in the valley to be able to broadcast local color with a color film chain Cactus and I went to WTOP in Washington and bought in 1965.

I have added a color "target 31" ID from the 60's at the bottom of the main web page at www.31alumni.com. It was originally in B&W hand painted by Howard Troutman. We had a "Coloring Contest" when we bought the used color film chain in 1965, and printed a "coloring book version" in the newspaper and asked viewers to color it and send it in. The winner got a color TV set and we would use the colorized version of the logo on the air, which we did. So, that's the history of the color target ID. My father liked the term "First In Huntsville" as 31 was the first station on the air, but it gave the impression we were #1 in the ratings, which was not always the case. My thanks to Val Ginter for sending it to me.

We also got color broadcasting started with the Network in 1965 (in 1967 we switched from ABC to the NBC Network) by passing color through the transmitter with a handful of parts Cactus was able to make and saved the great expense of buying an exciter made for color TV. So we added the film island and now could broadcast color movies (but the cost extra for the "print charge" for a color movie was $100, so we didn't order many in color unless they were a prime time blockbuster that was fully sponsored.

We could also do color slide commercials, but only live since we didn't have color tape recorders (those mentioned above came later). We went to audio cart machines and the new ones had "trip cues" to change the slides at the right spot. You recorded your audio commercial on an audio cart tape, then played it again, and as it played, you pressed a button on a box that inserted a tone on the other track of the cart tape. When it was played back on the switcher, each trip was sensed and did the same thing as pushing the slide change button on the dual drum RCA slide projector. This caused untold problems. The trip pulses were either two quick and nothing happened, or too long and the changed two slides at once, getting the whole commercial out of sequence on the air. This caused many problems and the short-term solution was to continue B&W video tape commercials or create the commercial on color sound-on-film. The problem with a film master was that it wore out pretty soon, got scratched and forever, the "film master" which was all the film commercials spliced together for the day, would break a splice and mess the break up. Also, the film leader got shorter and shorted until you were cutting into the spot or having to splice new leader again at the old splice and those often didn't hold. These were the problems for salesmen selling commercials that seemed to keep getting messed up on the air, but the technical crew was doing everything in our power to prevent it from happening. We even tried mylar tape splices on both sides of the film for strength, but we also succeeded in tearing up some projector gates, even more down time and trouble. We even found a TV station getting rid of their GPL (General Precision Labs) projectors and bought them and their parts and finally had enough parts, gates, spiders and such to keep the old monsters going a while longer. The Color Film Chain from WTOP we got, came with RCA projectors, so at long last, we retired the GPL projectors when we put in the RCA island. (I think we kept the B&W chain going for a while both during installation and operation as a backup and spare and had to use it more than a few times.)

We had "Artillery Punch" for a Christmas Party in the Studio. That mixture I learned in college. It was a combination of Hawaiian Punch and Pineapple Juice mixed with an ample supply of Vodka and Gin. I believe a half gallon of each.

The 10 p.m. news was a complete mess that night. It was because the TD, or Technical Director "punching the buttons" choosing cameras, film sources, running studio audio and keeping time for the show, imbibed also. I could not complain, I supplied the alcohol, but at all future parties, alcohol drinks were "Off Limits" to crew on duty for the 10 p.m. newscast. We supplied "soft drinks" for them and those who didn't imbibe alcohol. Things sure changed over the years on the alcohol at parties, ending with only beer and wine served.

Color Studio Camera arrive, the TK-42's. In the early 70's it was clear we had to continue our conversion to color and needed studio color to have full 100% color on the air all the time. It looked strange to see people on the news in B&W while the commercials and everything else was color. We had converted our processor to a HUGE color one in another new part of the building and sent Eric Eisgrau to school to learn how to mix chemicals and process color film (that at that time cost $25 a 100' roll). That was a lot of money at 2 ½ minutes out of a 100' roll if you shot 10 rolls a day, it was costing $250 plus the cost of color chemicals and it was a major outlay. Some stations only got 10% to 30% used out of what they shot in news each day. We tried for a minimum of 50% and even higher was rewarded. That meant you only turned the camera on when you knew it was stuff you wanted. Editing sound over B-roll footage ate up a lot more footage, of course, since you had the original film with the sound track and then shot silent film to run while the interview was going on. It was spliced in to the interview film, the magnetic sound track lifted off the original, and recorded back on the magnetic track of the silent footage. Quality was lost in the process, but it was accurate. To try to run an audio tape of the interview when the B-roll was running was asking for disaster. But it was how you came to look like a REAL TV station.

M.D. Smith, IV - '72 First News Consultants, On-air Changes, & Future

In 1972 we got our first consultants. It was the firm of McHugh and Hoffman. Phil McHugh was one of the owners and our consultant. We were being beaten 5 to 1 in the news ratings and he did both research and observations on what all we were doing wrong and open our eyes like they had never been opened before.

Several things resulted from this research. One, that M.D. IV was not well perceived as the weather man (I had been doing weather for over a year at 6 & 10 - being a pilot, I liked it and did it my way) so we needed a REAL meteorologist.

We hired "Wally Weather" as we called him around the station, but on the air, it was serious and he was Walton Jones, meteorologist. Wally still had a pleasant personality. We needed a high-powered anchor man and we heard about Sam Depino who was available in Mobile. When I talked to the GM at the station Sam had worked for he warned me, "Sam is a terrific anchor on the air and presents a fine news product, but behind the scenes, he's difficult for everybody to work with." I told the GM I appreciated his honesty, but I was in need of a top man that I could afford and Sam filled the bill, and I would just have to accept the other aspects of this man and get along with them.

As it turned out, Sam and I got along splendidly. I had long finished my Dale Carnegie training and would soon take the Management Seminar and then be qualified as an instructor for the course. I liked hobbies in woodworking, ceramics and flying and Sam liked and did the same ones. I still have several lamps I turned on my wood lathe from a trip Sam and I made to an old barn that was being torn down and we got some very good 6x6 beams of oak that day. Sam, however, did not get along so well with others on the staff. Sam was/is very smart and was a member of MENSA, the very high IQ organization. He often lacked tact, however, when expressing his critical ideas, objections and observations. I used to get VERY long reports from Sam, typed on green paper (second sheets). There was one occasion when a cameraman got into a huge argument with Sam and the cameraman said it's either Sam or me who is going. That time, it was the cameraman who left.

Ed Sisson was another hire about that same time, and Ed was smart and did a very good job reporting and anchoring some of the news shows (usually the 10 p.m.). Ed and Sam got into many "discussions" but Ed was one that appreciated Sam's depth of knowledge and experience so that they ultimately got along pretty well, even though some "discussions" they had could be heard all over the station. As I said, others came to like Sam less and less because he spoke his mind without holding anything back.

Our news ratings, with all we were doing with color film, starting to use a little bit of Color Video tape in the newscasts on a small basis, looking to the day we might go entirely tape, continued to grow. With the help of the McHugh-Hoffman consulting company and many different research projects over the years, Dan Whitsett made a chart that showed us originally being beaten 5 to 1, then 4 to 1, then 3 to 1, then 2 to 1, and finally, we WON the 6&10 p.m. news in total homes (that's what counted in those days) in 1974. Channel 19 responded and said over their two-way, "It looks like 'Gangbusters' (as they called us because of all the hard news we covered too) have won. Now we have to play catch up." So, they fired Hans Sitarz, got co-anchors (I think Tom Kennemer and Missy Ming, but I could be wrong on the first of their co-anchors) and they started to grow and took the #1 rating away from us. We got it back again in 1976 and the race was on for years to come.

I was impressed by the Frank N. Magid company and switched our research and consulting to them in 1975 and we stayed with them until I sold the station in 1999. I give them and their consultants a lot of credit for the ratings successes we scored over many years. Frank Magid is a genius and in several meetings with him, I developed great respect from the knowledge he learned about the TV business from his combined research projects from networks and TV stations. His stories about consulting for "Good Morning America," then "Today," then back to GMA were fascinating.

One of our news directors, Steve Ridge, went on to WKBW-TV in Buffalo and finally to Magid as a consultant. Today, he is the Vice President of Magid. Steve popularized the saying, "No Guts, No Glory" at our station. Many times, he wanted to try risky remotes, live shots or news series. I would get very uncomfortable and that's when he'd repeat his saying. We failed a few times with live shots and dramatic events, but the successes outweighed the failures by a large margin. Steve was right and a lot of reporters and other members of our news department benefited from his wisdom and drive.

Getting our 1,000-foot tower built in 1976 (to match 19's new tall tower) was a good idea and we needed to get a better signal into the shoals, which was becoming ever more important in the rating books. We could not afford an RCA transmitter, so Cactus negotiated again with George Townsend (we put the first Townsend amplifier in the station when we built our 400' tower that was removed when the 1,000 foot was built) to get a 50kw transmitter using Varian Klystrons. The aural used the same tube, but at 50% power (some years later the FCC let stations go to 10% of visual power for the aural transmitter). We could multiplex out of either transmitter if there was a failure of one of the tubes. Cactus and the engineering department built a LOT of the plumbing, custom patch panels, spare pumps and heat exchangers and we had a lot of backup. Thanks to Cactus and the engineering "boys," we had the lowest costs and smallest amount of lost air time of any station in the country compared to the perfect product we put on the air.

The "tower and power" and our #1 news ratings, helped us get the ABC network in 1977, which was then, the #1 Network in the country. It was a very big event. We brought the entire cast of "Carter Country" to Huntsville for our change-over party at the Hilton and all our best sponsors and VIP's were invited. The custom jingle package we had made was the same as the "Still The One" theme music, but included our call letters and channel number.

Years past and we continued to wage the "News Wars" with channel 19. Channel 48 was not a factor in the news ratings until the mid-90's. At one time, we grew to a staff of 175 people. For a market ranked around 90, this was over double what most stations, including both local Huntsville stations, of even markets near 50 employed. But we had news 24 hours a day (LIVE reports on every hour) and 7 days a week. We had many special remotes and events we sponsored every year. In 1995, we had double the news cars and live trucks as well as five manned bureaus in the coverage area, including Fayetteville.

Side note here. The Lower Lot. It was originally just a small concrete pad near the bottom of the big 1,000-foot tower, but we enlarged it to use for parking and built about 90 wood steps going up the mountain to the edge of the "top" parking lot. As you can imagine, it was not a popular place for employees to park and walk up all those stairs (no matter how good it was for you). But parking was overflowing on the top lot and we needed to get some employees to park on the lower lot to leave room for clients, guests and others who needed more than the one visitor parking space we had. So, the "Lower Lot Sweepstakes" was born. Here's how it worked. We had a cash prize for the person whose name was drawn each month. You got a "name slip" each day that you walked the steps and went by the receptionist and registered. The more times you parked down there, the more slips you got and thus increased your chances to win. It did work to a large degree and I was happy to pay the cash prize (somewhere between $50 and $100—if someone wants to correct me) and a round of applause at our regular Employee of the Month meetings, where we also announced other company business and had some limited discussions. I do remember a consistent winner. It was Karen Petersen. She walked the steps consistently more than other employees, but sometimes even won with less slips than other regulars who walked the steps.

We did a commercial for News in early 1995 showing our Fleet of News cars and trucks. Our four news anchors at the time, Don Phelps, Heather Burns, Cliff Hill and Liz Hurley did the on-camera parts. We showed our entire fleet of marked news cars in the lower parking lot and it was impressive. Our big Live Satellite Truck (LIVE STAR 31) was on the way and would arrive in July of '95. (Just a week before Channel 19's smaller live truck came to town, and THEY claimed to have the first live truck). We put a stop to that.

In 1999 we decided to sell the station from an offer that was too good to refuse (and facing growing costs from expanding HiWAAY and the facing of High Definition cost on the near horizon). This was just before the dot.com bubble burst, 9-11 and other factors that made the timing perfect to sell a local station.

People have asked me if I miss the business. The answer is yes and no. Yes, I miss the excitement of live TV, breaking news events and planned live remotes. The complete news casts on location from places like Opryland and on the Tennessee River on a barge were amazing. Playing the ratings games, news and weather wars, and out-promoting the competition (or beating them to the punch on major breaking stories) were all the fun and exciting parts of the business. Dealing with the disgruntled public (I always allowed direct access without having to go through a "screener"), some of the personalities of anchor people who got to believing the promotion about how great they were, lawsuits, and a thousand other urgent emergencies (including lightning strikes that could cost us thousands of dollars) are not missed a bit.

I have a thousand "small stories" of individuals who worked for 31 over the years and from some of the Christmas Parties for staff only in the studio. (We had a very strange 10 p.m. newscast one December after the party when the director, audio op and switcher had been imbibing). It happened more than once, until we forbid anyone working that night from having anything alcoholic to drink.

We valued creativity and let a lot of people have more reins that they could get anywhere else. Lew Koch, production manager (later Operations Manager) understood this and that was the birth of some wild commercials (many of which never made it to air). We even had a few promos we ran just a couple of times, late at night so we could enter them in the Addy Awards because they were so creative. But they were in bad taste, took direct shots at the competition or simply didn't get the core message across, so we never ran them for a length of time. Most of them took first place in competition.

The "WAAY TOO LATE" crew and program that came on very late on Saturday nights, was extremely creative and sometimes even got me in hot water with the public (yes, at 1:30 a.m. a few people were watching who got offended by the weird or gutter humor). But these same people, enjoyed the atmosphere and stayed to produce great shows, commercials, promos and programs. That goes for sports and news as well.

I wish there was more than a few hundred aging ¾" tapes in my storage office to show for all those years of great things Channel 31 accomplished (only because of the people). This bit of recollections and the web site which host it currently (2003) are an attempt to have these stories and great memories live a bit longer. Perhaps someone will commit these stories and web pages to hard storage media someday. The View I had looking out of my office for many years. I never changed the orange shag carpet because to move my desk, two computers and all that audio and video hookup was just too much trouble. When I left to move to the home office on Clinton Street, and Redge Swing took over as G.M., that was the first thing he changed. I don't blame him.

Did you ever site in that same black leather chair that you see in my office looking out towards my Administration Manager's office? That was my view every day from behind the desk.

Hank Price – Production Manager's Stories, Santa, Game Host, Promotion

Several Updates added

In 1975 I was the Production Manager *[Editor's note: and Hank also did double duty as announcer for Dialing for Dollars cut-ins during the afternoon- see stories near bottom]* and Bobby Gleason was the Film Editor.

Bobby kept coming to me asking for a raise. Claimed that he had not had a raise in three years and could no longer live on $93.00 a week.

I asked M.D. IV a few times and kept getting turned down. M.D. felt $93.00 was a fair salary and that was that. Finally, in desperation, Bobby asked if he could "go see Mr. Smith." I of course agreed, was in fact relieved. Although I warned Bobby: "You're not going to have any better luck than I did."

Much to my surprise, Bobby came back from the meeting elated and full of confidence. "It went great!" he said. "I explained to Mr. Smith how I hadn't had a raise in three years and why I needed more money. He was very understanding. He asked me to go make a list of all my expenses and bring it back for us to go over. He said he would help me put together a budget, and if it turned out I really needed more money he would give me a raise!"

For the next few days Bobby worked diligently to gather all his bills and other expenses. Finally, with an air of excitement and expectation, he went to see M.D.

About 10 minutes later Bobby was back in my office, devastated and ready to quit. "What happened?" I asked. "What happened!" Bobby replied, "What happened was that Mr. Smith looked at my list. The first item was my car payment, $122.00 a month. 'Here's the problem,' he said. 'You can't afford a $122.00 a month car payment on what you make. You need to sell that car.'"

Bobby sold his car.

Updates added about Santa Show, Coffee Break and More from Hank.

SANTA SHOW

I was thrilled to get to do the Santa Show for one reason and one reason only: $15.00 a week talent fee. That was a lot of money to me at the time and well worth putting on the smelly suit and yellowing beard that had obviously seen better days.

My first week on the show I found out exactly why the suit was in such bad shape. We were going along as normal. In my deepest voice I said "Tell me Miss Merry Christmas. Who's our next little girl?" as I looked down at a terrified four-year-old who obviously wanted to be anywhere but there. Half pulling, half pushing the child on my lap, Miss Merry Christmas said, "This is Judy."

"What would you like for Christmas Judy?" I asked as kindly as possible. Judy did not answer. Instead, she looked up at me with horror, squeezed her lips together and turned red. Suddenly my entire left leg became warm. I jumped back as Judy tumbled off my lap, but too late. Just another chapter in the suit.

Another time, my wife Maria brought our two-year-old son, Pepper, to be on the show. Pepper took one look at me, became frightened and ran out of the studio. Little did he know that scary Santa was his father.

DIALING FOR DOLLARS

M.D. decided our afternoon ratings would be enhanced if we started doing Dialing for Dollars. We got a picture of a set someplace and Dave Troutman went to work building it. The wheel was an old bicycle wheel I found at a neighbor's house. Dave took the tire off, put a cover on top and bolted it to an easel. Worked great.

Because the Dialing segments did not pay a talent fee, there was no permanent host. I seemed to do it most of the time, but it could easily be Mike, Carl or anyone else who happened to be around and had a coat that day. Occasionally we would all be tied up and I would have to grab Dean Abla to do it, or even the weather guy.

As I recall, we cut a phone book up and put the names and numbers in a bin. The host would call a number and if the person was watching and could answer the question, they would win a prize. Cutting up that phone book was so much trouble that we wouldn't bother to cut up the new books when they came out every year. Occasionally someone would win, which was always a shock.

COFFEE BREAK

Coffee Break was an amazing show. Hosted by Johnny Evans and Mary Beaton, it ran every weekday morning at 9 a.m. The show was directed by any number of people, depending on who happened to be available. One of the things I would have to do every morning was find a director and a camera crew. The crew usually consisted of Dave Troutman and Bobby Gleason, both of whom I would have to force into the studio because they hated running camera.

Mary would always arrive early, greet the guests, and write down all the important information. Johnny, on the other hand, never left his office until he heard the opening music. He would then slowly saunter into the studio, sit down next to Mary, and put his microphone on just as the camera went hot. Without fail, every day, five days a week,

Johnny would start the show by saying: "Tell me Mary, who are our guests today." Johnny always used those words because he had no idea who the guests were or what questions he was supposed to ask. Sweet Mary would always tell the audience (and Johnny) all about the guests.

One continual problem with the show was that the set was raised and open under the desk. Anytime a woman with a short skirt sat down and Bob Turner or Dave Troutman happened to be running camera #1, we would get an embarrassing close up. I would start yelling over the intercom to fix the shot until the camera slowly got back in place.

We also had a problem with Johnny trying to pinch Mary. She usually squirmed out of the way but was not always fast enough.

And of course, the show always ended the same way. Johnny would get up, take his mic off and leave as soon as the music started, even though he was still on camera. End of another episode of great television.

THE FIRST HONDA COMMERCIAL

Sometime around 1973 one of the local car dealers added the new Honda Civic car to its product line. I was assigned to produce the commercial so Bob Turner and I threw the film equipment into his old (very old) beat up Pontiac. The Pontiac was distinctive because it had four different size used tires on it–all that Bob could afford. Bob and I drove up in front of the dealership, went inside and said we were there from WAAY to pick up a Honda car. We were given keys and went out back where when found this cute little orange car. As we drove around and left the lot, I noticed a group of salespeople gathered around Bob's car. It was dieseling, pouring black smoke out the back, and shaking all over. A few hours later, when we returned, a very hostile car salesman showed us where they had towed Bob's car to.

As I recall, Bob was out driving one day when the car died in the middle of a busy street and would not start. In typical Bob Turner fashion, he got out and walked away. Never saw the car again.

THE COMPANY PARTY

My #1 all-time WAAY story happened at the company picnic. We were all invited down to the "big house" on a Saturday afternoon. Mr. Smith III had soft drinks, beer, hamburgers and hot dogs. After the party had been going on for several hours Mr. Smith asked everyone to gather round so he could show us a dog trick. He said he was going to throw a can of beer in the pool and his dog would retrieve it. Mr. Smith threw the can and motioned to the dog. Nothing happened. The dog just sat there. Mr. Smith tried everything. No reaction from the dog. Suddenly people started whispering to each other. We were all convinced the dog would not go in the pool because the employees had been in it.

But that's not the main part of the story. Much later in the day Joe Garerri, who had had way too much beer to drink, came up to me and said "I've got to tell you something. You won't believe it!"

"What Joe?"

"A few minutes ago, I was desperate to go to the bathroom. There was a long line at the pool house so I went into the main house. A long line there too. I was about to bust. Had to do something, so I went upstairs and found a bathroom. I think it had to belong to Mr. and Mrs. Smith. You'll never believe what they had in their bathroom right next to the toilet—a water fountain!" *[Editor - this was the bidet.]*

[Editor's note: This update from Hank after he saw the 1981 ratings sheet posted on the web.]

Hi M.D.,

Just read the posting about 1981 and winning. I don't know if I ever told you this, but my first impression of WAAY was that the facility wasn't much to look at, but the product was the best I'd ever seen. Now, having run four television stations of my own, I'd like you to know that at each one I've told people about the winning attitude and great product we had at WAAY in the 1970's. I love to bring up the fact that you couldn't tell anything about the on-air product from touring the facility.

Over the years I've spent a lot of money on buildings and equipment (over 5 million in the 2 1/2 years I've been back in North Carolina). Each time we make a big expenditure I remind people that there is no magic bullet. Sat trucks and helicopters do not make a winning station. Product, marketing and branding all matter, but in the end the desire to do a superior job and WIN always makes the difference.

Part 4

M.D. Smith, IV 1994 Staff Meeting – Insider's Info Only

Ten years ago, I gave this talk at two staff meetings (one at night for news department and the other during the day for the daytime staff to attend). I thought it would be interesting and somewhat telling of the cautions and warnings I made then which have come to past. Here is it.

[Snacks to be served 15 minutes before each meeting.] 3-22-94

[DRAW for DOOR Prizes and make awards.] [6:46 pm 3-21-94]

You will all hear the same words I speak. How you interpret the words will be different. What you think I mean will be different. What you believe is behind some of the words will also be interpreted differently. But if there is just one common conclusion I hope you will draw, it's this: We are in this boat together. Whether IT wins the boat race or loses...or even sinks, is up to all of us. It's up to me and each and every one of you. It's for that reason I am here NOW.

Linda Spalla, G.M. at Ch. 19 recently told a person in the media, and it has been widely repeated since then..."WHNT's MAIN GOAL is to beat WAAY-TV into a pulp." To quote from "Home Improvement," No, Linda, I DON'T THINK SO. But that seems to be their mission. OUR...MISSION STATEMENT:

Mission Statement of Channel 31 - WAAY-TV

To produce RATINGS and REVENUE in increasing amounts to support a healthy growth of the station and the employees.

SUPPORT GOALS:

RATINGS come from satisfied viewers that view our programming and enjoy what they watch more than other stations. The newscasts are compelling and so interesting they call others to the set to see. A solid promotion and advertising campaign adds to both the perception and real benefits our unique strengths give to our viewers. Technical excellence aids this satisfaction.

REVENUE is generated from advertising sales and must always exceed expenses. Clients who see results from their advertising, reaching more people for less dollars will always pay top price to be on our station.

NEWS: To fight to remain the station more people in the Tennessee Valley turn to for News, Sports, Weather & Special Event programming. We must continue to build, innovate, grow and perfect our methods and style. We have to stay informed of the wishes and needs of our largest audience segments. We have to be VIEWER focused.

IMAGE: That we are always available, 24 hours a day, to satisfy the largest segment of needs for information, while providing entertainment, sometimes blending entertainment with information. We are the most alert, capable and responsive news organization exceeding expectations of a station our size.

ENGINEERING: To provide support for all technical aspects of our station. To keep all operating equipment at the best state of repair so as to be as reliable as new. To innovate with new technology, while keeping costs as low as possible.

SALES: To produce revenues in increasing amounts meeting or exceeding operating costs adjusted on a yearly basis. To provide customer results that exceed expectations and to continue to find innovative ways to blend their needs with what the station can deliver.

PRODUCTION: To provide superior on-air productions with highly trained and skilled people. To generate commercial and station recorded messages that are outstanding in both presentation and results generated. To present a total on-air image that is clearly superior to the competition.

STAFF: To remember that every job is very important or it would not exist. To look for better ways to do every single job on a continuous basis. To be open to change as ways to keep the work interesting and benefitting the total effort of the station every day. We must work as teams within departments and with other departments as well.

PEOPLE DEVELOPMENT: Recognizing that each staff member has needs for career advancement and financial security. To stimulate and assist every person to grow and become skilled at new areas of work on a continuing basis. To aggressively pay our staff for innovative, motivated and improved work. To have deserving staff members enjoy the fruits of our combined successes.

LEADERSHIP: To develop increasing numbers of leaders and trainers. To know that leadership is the ability to enlist the willing cooperation of others to achieve desired results.

++++++++++++++++++++++++++++++++++++

1. CHANGE. Story of the lily pads in the lake, doubling until on the 30th day it exactly fills the lake. How much of the lake did it occupy on the 29th day? (Half) Regarding change, we are on the 29th day. It is easily possible that three years from now, society and our economy and certainly our industry (TV) could undergo as much change as it has in the last 10 years. Three years from TODAY, we might look back at the TV business and long for the good, old days when things moved as such a slow pace. Competition was manageable. Our abilities could be truly satisfactory with less than half the efforts required in 1997.

In the 70's, Organizations said they could satisfactorily get by with:

No Change --- 60 %

Sporadic, Incremental change --- 35 %

Continuous, Overlapping change --- 5 %

In the 90's, Organizations said they could satisfactorily get by with:

No Change --- 1 %

```
Sporadic, Incremental change --- 24 %

Continuous, Overlapping change --- 75 %
```

Even solutions to today's problems, have a much shorter "Shelf-Life" than ever before. Relate story from page 45 and 46 from "Managing at the Speed of Change" (Miners in the Dark, but one had a spare, stick match he always carried in a container). Today's solution may not work tomorrow.

The way to be successful in the 90's is to Change the Way WE Handle CHANGE. It takes flexible and resilient people to understand and use to their advantage the changes in the workplace, industry and certainly, their own lives. It can be stressful, but it doesn't have to be.

The way to prevent becoming a victim of it, is to be involved with it. Embrace it. Be a part of it.

2. IDEAS. There are many cases of brilliant ideas poorly executed and the result was a total failure. How do we improve the execution of new ideas? We GET YOUR input on them. We need ideas from staff on how to accomplish a lot of things better, easier, different. Name Tag ideas that have come from staff are going to result in a very nice customized logo ID that you may wear proudly off the job if you wish.

3. Cover ratings information on newscasts. We were happy with the results...but we need to not be lulled into a false sense of security. Ratings are like a big battleship changing course...when you notice it's off course, it's way too late to start making corrections. It takes a LONG time with a lot of pressure to change it back. We can't wait until we see bad news in the ratings. We have to continue to BEAR down NOW...while we still have a competitive advantage.

4. LOSE the CHAFF. In separating the good wheat from what is not, you keep shuffling the mix around until the chaff floats to the top and is discarded. We have to do the same with functions, procedures, policies and even some people. In the kind of environment we are working toward at 31, we have to have everyone pulling their weight.

5. COMMUNICATION. Must be more effective and more people oriented, but goal driven as well. IWWCW...In What Ways Can We… accomplish more with same or less. "People forget the words you use, but they never forget how you made them feel." Our body language, tone and inflection of voice transmits much more than the actual words themselves. But some words transmit powerful feelings, especially negative ones. Better leave angry words unsaid. People also understand different things than were actually said. Cactus told me of two good ole boys walking in front of the Old Time Hardware store on the square in Scottsboro on First Monday...and there was a sign in the window of the hardware store...it read: "CAST IRON SINKS." So, this one guy said to the other one..."Heck, any darn fool knows that!" Other people's attitudes at the time are important, too. You say to a woman who is having a really BAD day..."That's a pretty dress you have on today." and she retorts..."What the devil was wrong with the one I had on yesterday?"

Teamwork is vitally important to our efforts and reaching our goals. We can't allow poor communications and communicating problems to affect us. There are been quite a few hurt feelings and anger of late, due to people saying one thing and it gets taken to mean another thing. Talk things out...and not when you are angry. God made you with TWO ears and just ONE mouth. You should LISTEN TWICE as much as you TALK. Do you remember someone in your past who REALLY, REALLY listened to you? How did it make you FEEL? That's the very important point here.

6. TEAMWORK. Two plus two is greater than four. Teams that are empowered, get faster and better results.

6A. Don't be afraid to fail. It's OK to make a mistake (as long as it doesn't sink the ship.) READ "DON'T BE AFRIAD."

7. Remember the doubling of a penny a day for 30 days. First day 1 cent...next day 2 cents...next day 4 cents...etc. Finally, how much at the end of 30 days? Anyone know? A millionaire. ($ 5,368,709.00) but for the first two weeks you would only have earned a couple hundred dollars. Don't quit!

Leadership in the 90's is like putting yourself in a foxhole with several of your employees. All your lives depend on your best combined actions. Leaders want to hear what the others think. What their ideas are. As a member, each person knows that their ideas will be accepted in the spirit of cooperation and looking for the best combined answer(s). They are not intimidated by the leader but see that person as there to help just as much as everybody else. But remember what writer H.L. Mencken said, "For every complex problem in the world, there exists a simple solution–which is almost always wrong." You first solution is often NOT the best solution. Keep on going from there. Keep on finding better solutions, even if you have to implement the first one right now. Stopping just short of GOALS and accomplishments is for the competition to do. We don't give up around here.

FUN: There should always be an element of fun and enjoyment about the job, even if it is stressful at times. Everyone should have parts of their job that makes it fun to come to work each day. If not... you should discuss it with your dept. head or with me.

(THE PROBLEM SOLVING FORM...4-part form large and small sets)

1. **What is the problem?**

2. **What are all the causes of the problem?**

3. **What and all the possible solutions of the problem?**

4. **What is the BEST solution to the problem?**

++++++++++++++++++++++++++++++++++++

Use the suggestion box if you wish to be anonymous or take to your dept. head.

* TRAINING. To be all that you can be and more, we ALL must seek additional training. Ask for it. Seek it. Find it. Let us help. "You have to have a high attitude before you can reach a high ALTITUDE." That's an old Navy Pilot saying.

We are in this boat together. Whether IT wins the boat race or loses...or even sinks, is up to all of us. It's up to me and each and every one of you. My commitment is to steer this ship at full steam, full course ahead. I am proud to have each one of you aboard.

Building The Motivated Team (Dept. Head Meeting)

by M.D. Smith, IV - - - March 15, 1994

In a nutshell, we need to motivate the staff as a TEAM and get coordinated teamwork to reach goals. Our goals are to produce a better product for the viewer, primarily in news, and have them choose us over the competition. Our advertisers must get good value and response from their commercials so they will continue to give us the big share of their budgets.

We have to instill the value and drive that the small, free enterprise individuals have in competition today. Our staff has to understand our goals. They have to BELIEVE in our goals and see that getting to the goals, benefits THEM. If they don't understand "what's in it for me?" we cannot motivate them to do anything. They have to feel and be involved in every aspect of getting better and of solving problems (opportunities for improvement; OFI's).

Businesses that are succeeding today are doing things in a very simple way, doing them regularly, and never neglecting to do them. (Says William Hesketh Lever)

The owners of small businesses that are succeeding believe they are creating value for customers, employees and of course, themselves. The entrepreneur lives and dies on his answers to "what" and "how"– and they are always connected.

"Strategy" is a matter of survival. The purpose of creating a company "culture" is not more wall plaques; the ONLY purpose is to get everyone focused on those few values that will give us the competitive advantage to make the business strategy come true. When the values directly support the product/market strategy, watch out. It's the most powerful way known to energize a group of individuals to achieve a common purpose. (Quotes from article by Larry Farrell)

Walt Disney was obsessed with making products that customers will buy. (The organization still is). They are committed today to providing "exceptional" service and treatment of their "guests." They strive to ALWAYS deliver more than was promised and deliver more than is expected (which can be quite high) to each and every guest.

Mr. Toyoda of Toyota realized as long ago as 1947 that the secret to running a successful business was to revive the craftsman in all of us which means to run the business from left to right, not top to bottom. This is the Disney and the Toyoda way, the old-fashioned, entrepreneurial way.

Empowerment that we talk about is really having a sense of urgency to innovate so that innovation becomes a necessity and everyone has the freedom to act, and act quickly.

We cannot wait for disaster to strike in the ratings. We have to start today. We all have to have that feeling of the NECESSITY to do it better. We have to instill that same feeling in each and every employee –the necessity to do it better–and granting the freedom to do it faster.

Sam Walton (of Walmart) said: "Our folks don't expect something for nothing . . . they want to win so badly, they just go out there and do it."

When you think about it, it's pretty darn tough to beat people who love what they do and are darn good at it. How do we create self-inspired employees? Walton knew many of the ways. From his whoop-it-up Wal-Mart cheers and profit sharing for cashiers to steely intolerance of slacking off and losing ground to a competitor, Walton created a company full of people who love to win and hate to lose. In doing so, he tapped the most powerful drive in human behavior: self-interest. People tend to do things that bring them positive consequences and stop doing things that either bring them NO consequences or negative consequences.

The successful entrepreneurs face consequences of their behavior and actions, positive or negative, every day. This almost never happens in a bureaucracy. With neither positives nor negatives, managers and workers float in perpetual limbo, oblivious to the fortunes or misfortunes of the company and their part in it.

AM I STRIKING SOME FAMILIAR CORDS? IS THIS MUSIC WE HEAR FROM OUR PEOPLE SOME (OR MUCH) OF THE TIME?

Our people HAVE to hear when they do things right. In a friendly but unflinching way, they have to know when they do things wrong. We have to help them correct the problems. They must look at problems and mistakes as REAL OFI's (Opportunities for Improvement).

We managers MUST spend about three-fourths (3/4) of our work week doing two things. One, motivate each person that works for us to WANT to REALLY SUCCEED and enjoy the fruits of their success (both in compliments and acclaim as well as tangible rewards). The other is to help build them to be more successful, once we kindle the desire. But remember, it's like writing in the snow, you have to keep going over and over the same lines if you want your writing to remain legible. Leadership and motivation is just like that. It never stops. It is never the same. There is no rule book you can follow day in and day out. YOU have to innovate, just like you are expecting your people to do, in your ways to motivate them to peak achievement. The OTHER one-fourth of your time should be spent doing everything else your job requires. If you don't have enough time in that one-fourth, you need to quickly work on ABCDE...Always Be Combining, Delegating and Eliminating.

As Soichiro Honda once said, "I want to touch and hold a better piston, not watch another concept presentation."

If the bureaucracy is holding us back, eliminate it and start over. WE want high purpose, absolute focus on customer and product, a lot of action, and self-inspired people. It's called enterprise.

This is part of the continuation.

M.D. Smith, IV March 15, 1994

MDS July 1995 Staff Meeting – Confidential, until now

Nine Years ago, I gave this talk and it was after Ch. 19's new News Director Craig Jahelka (now Vice President/General Manager KERO-TV Bakersfield) took over and Dan Satterfield went on the air. They were definitely out-promoting 31 and the research proved what was beginning in 1993 and 1994 was very effective.

Photo above with Linda Spalla and me was a year later in 1996 when 31 and 19 were definitely in the "News Wars."

Meeting notes for Monday, July 10, 1995 with staff.

(Show some of Ch. 19's 6 p.m. Thurs. show on Sat Truck.)

We have a significant problem. The attitude study just done by Magid shows almost the same ratio of preference as did Thursday night, May 18th during the storms and tornadoes, in the Nielsen rating book. (SHOW PIE CHART)

31 is the preferred news station by 27%, while 19 is preferred by 45% and channel 48 by only 9%. This is very bad news, but it can change if we react quickly before it gets etched in stone and the rating books.

We learned much, much more from this study, but in very simple words, there is not a single problem that we have or a single reason 19 has done so well, but the bottom line is, they do good TV News and Weather, and they understand the "SHOW" and presentation of it. They market well, it is unified and complete.

They have been growing, trying new things, adding new features, elements and ideas to their programs. We, on the other hand are seen as good, but much the same as 3-4 years ago. We may be a Cadillac product, but there is rust on the fenders and scratches and dirt on the car and the engine, from just running at idle speeds, does not seem to perform like it once did (IN THE EYES OF THE VIEWER). 19 is seen as the station that has been getting better, while 31 and 48 are seen as not as good as they used to be, by a number of people.

Area differences...shoals WHNT=55%, in DCU 39%, Lincoln 64%, east DMA=45%WAAY = 19% shoals, 7% Lincoln co. and 32% in Hsv for preference levels.

Elements of decision-making process...Rated for strongest reason for choosing one station over another. (All loyalist say same thing. The AMOUNT of news the station COVERS from YOUR area = 83%)

Newscast imagery: How THEY ranked US (31) in most mentions

Covers news 24 hours a day.... AAY=42%, HNT=29%

Best coverage of news across the Tenn Valley AAY=30%, HNT 44%

Most live coverage AAY= 28% HNT = 43%

Most resources AAY=28%, HNT=41%

Best Job of covering news from where you live AAY=28%, HNT=48% AFF=9%

Best Coverage of shoals AAY=28%, HNT = 37%

First with late breaking news.... AAY=28%, HNT=44%

Best coverage of Hsv AAY=27% HNT 48%

Most enjoyable to watch AAY 27%, HNT=49%

Most knowledgeable news anchors AAY=27%, HNT=40%

Works to improve the local community AAY=25%, HNT=37%

Makes the news easy to understand AAY=25%, HNT =46%

News anchors who really care about the news they report AAY=24%, HNT=43%

Understands what's important to you AAY=24%, HNT=45%

Decatur news AAY=21%, HNT=30% AFF=15%(higher than their 9% base)

Better than used to be AAY=18%, HNT=44%

Gotten worse than used to be... WAAY=11%, HNT=6%, AFF=21%

Then consider OUR loyalists vs. others...do ours name us for all those elements...70% is a desired level from a station's own loyalists.

We were good, but our loyalists say only 64% for shoals. Coverage from Huntsville, we are also low at 66%....

Importance of Weather.... 84% VERY Important, 14% somewhat.... 3% NONE

How Interested are you? In addition to covering the local news....(that's #1)

1. Good news that is happening in the community 76%

2. Stories that reflect the value of the family 67%

3. Stories about health and medicine.... 60%

4. Summary of the day's top national stories 59%

5. Local people/places of interest 51%

6. Report about consumer issues 35%

7. Religious stories 32%

===

There was a time we were criticized when we made a big deal out of weather...that we were scaring people...and we were a dominant #1 in weather....now we don't get criticized about that...but channel 19 does do it...and look who's #1 on the weather. We should go wall to wall on everything that's any kind of big weather story and do coverage in ways the competition has not thought of....

Night of Wed. July 5th, 10pm...We did lead with weather, but 19 had people in Arab, Live heading for shelters and cover....and our 3 packs of weather...lead with weather of evening...story of severe weather...real or perceived...better technology? Actually, a bit of both....but we have TWO claimed Weather Experts...but there was nothing from Bob and Gary, from a Marketing perspective? Why were they not included in the story? We did story on mobile home story and shingles and could not get it carried off....but that was another story that could have been tied together....it could have been "big story, team coverage" and all this depth, but we didn't do and didn't sell it well at all.

The big story needs to be blown out of proportion...no lies, but really blow it up and sell it. Same whenever weather starts to turn ugly.

Shows look like it is just stories that are stacked together, not a program with planning, sidebars, and stories that make sense with the whole show theme. One thing, Mark is back on the assignment desk and things are working better. Do the producers know in the morning what today's show is going to be? Seems that they don't....they are told what the stories are for today...and they just stack them.... Is interaction taking place? Dot says there is a bit of that going on, but not enough.

"NO EXCUSES?"

We have 3 or 4 separate weathercast on the same station...so how can anyone latch onto the way we do weather and put their arms around it....Dan's weather is defined....and everyone does it that way....WE have to be consistent on the weather. You don't by a can of coke, and depending on who is working that day, the Coke taste different....The viewers HAVE to know what to expect consistently and it has to be good...clear...and easy to understand.

Dan...LOCAL...where I live...easy to understand...it affects me where I live...I can understand it...

==============================

COVER THE NEWS FROM WHERE IS LIVE...DO IT AGGRESSIVELY, CONSISTENTLY AND WITH FLARE AND URGENCY.

Overview...HNT and Dan cares...that's the message. They're out there when it counts...This is a desperate call to arms...ratings are going to be a disaster...probably even before we could do anything much about it...but we have to pour on the coal, all the burners...and do everything we can to stop it. Everyone here at 31 has to get more serious about what they are going to do.

We have to be much smarter with LiveStar than anyone...get our money's worth, which will mean spending more money with sat. Time and keeping the truck on the road...using it smartly and strategically will be the key.

We have to live up to our "COVERAGE YOU CAN COUNT ON" and really show that we are the News and Weather station with more of both, more often, more live and anything that connotes we are the best at it. Whenever LIVE news airs...make sure it's taped. Then in later newscasts...show how we broke in and did the coverage…and show a capsule of what we did.

If we promote a big story everybody has and we only promote in generally, the audience will go to their favorite station to see the story, assuming everyone will have it...and won't tune to US. So, don't tease, "A BIG FIRE DOWNTOWN, AT 5 WE WILL TELL YOU WHERE IT WAS!" That makes everybody mad that you are trying to keep them in the dark. Almost as bad, "A BIG FIRE DOWNTOWN AT CITY HALL, WATCH US TONIGHT FOR DETAILS." No, again, everybody has it and they will watch their favorite station. BUT if we say, "A BIG FIRE DOWNTOWN AT CITY HALL AND WE TALKED WITH PEOPLE TRAPPED IN THE ELEVATOR DURING THE ORDEAL. HEAR THEIR SCARY STORY AND HOW THEY REMAINED CALM TONIGHT AT 5!" Now that's going to hook some people that might have gone elsewhere.

Even in bad weather, we have to drive them to our station, or they will go to their favorite station...saying something like, "We are going LIVE to the bad weather area and talk with a Florence resident who has a special story to tell about being prepared."

We have to be so compelling with all we are doing and about every element of news and weather...that what we promise about our coverage, we do...and we tell them we are doing it. We have to keep our severe weather coverage or big event coverage so compelling and constantly teasing what we have coming up next, that people can't afford to leave us for the competition. We have to keep them GLUED to the set on US.

[side note: Big Red vs. The Great White]

We need 31 LiveStar bumper sticker and mini stickers to give out whenever LiveStar makes an appearance at any news story, location or event. Have hats to give out to participants and key personnel like Mayor and Sheriff and Police chief of various cities we visit. Have them on for a flash for interviews, too.

We are not properly EXPLOITING our LiveStar or our news the way we should. Make every story, a big story element of our news. Tie every element of the news together as part of the big picture of that day's show. We need to have "big story" meetings long before the newscast is cast in stone from story content and presentation and plan for a "whole show theme" for every cast...and change it when the big story breaking demands it. We are not in the "history" business.

To the audience out there...19 is a fresh, new product, making advances and strides every day. They are exciting. Something is always happening there. At 31, we are the same old car we were 4 years ago...not nearly as new, getting scratches and dents. We were a great car when we were new a few years ago, but nothing much has changed in their eyes and we are "just good" but not exciting and on the edge. It may be a lot of perception as well as what we actually do...but perception is reality. The research tells us it is firmly based in performance, leadership with news and weather, being first, proving it, showing it, advertising it, promoting it within newscasts and every element (including even the 5:30 a.m. morning show)...at 19 "They ALL get it!" They understand to tie in the YOU...YOU...YOU into everything they say and do. It matters to those folks out there. It matters a LOT! We have to understand and be more "UP CLOSE AND PERSONAL" than they ever thought about being. We have to be more "YOU-oriented" than ever before.

==

Bottom line

--Resolve to Win

--Vision

--Blueprint

--Knowledge/Ability

--Work Ethic/commitment

It's going to take…IMAGINATION, BURNING DESIRE, CREATIVITY AND DETERMINATION to make us winners again. Every single person has to literally "catch fire" with the leaders (who better be on fire already) to take us out of this downward plunge...or else.

It's GOT to HAPPEN.

Use your imagination in ways you have not before. Go out of this meeting with fire, resolve, determination, and a commitment to use your creativity and...your...

(Clip from song....Imagination). You. My commitment is to steer this ship at full steam, full course ahead. I am proud to have each one of you aboard.

Part 5

Running the TV Station in 1995

By M.D. Smith, IV

A previous part of this story with the same title took place in 1964. Thirty-one years later, running a TV station was vastly different. My father had died in 1985 from lung cancer, and I was completely on my own, with gentle suggestions from my mother who still kept the books, wrote the checks and was the majority stock holder. She died in 1997, and I chose to sell the station in 1999. I'll give you my reasons for the sale at the end of this story.

I am glad she lived to see the largest the TV station would ever get in 1995 with one hundred and seventy-five employees. If I had not lived and worked through every single one of the years from 1963 to this 1995 time period, I would find it hard to believe. And for the most part, growing audiences, ratings and income continued to grow as well. It was only in the years after 1995 that TV audiences, in general, began a sharp decline, spending time with other media, rented movies, special cable channels and soon the internet.

March of 1995 saw the birth of our internet company. We had bought a T-1 line from the TV station to a company in town for our use and decided to use the extra band-width to handle some "dial-in" telephone modems and make enough money to pay for our T-1 line. In March of 1995, we joined with "Information Highway" with a newly formed separate company named, **"HiWAAY Information Services."** We liked to use the word "WAAY" in anything we did, and Mark Derrick our resident System Manager and computer guru suggested the name. We added subscribers faster than we could buy new modems and get additional telephone lines run to the computer room.

This was a very big year, 1995, for additional tools for the News Department.

July 7, 1995 - LIVESTAR 31 arrives in town. This is Channel 31's first Satellite Uplink Truck and Mobile Studio. It cost three-quarters of a million dollars. It is a monster, and bigger than the one Channel 19 has purchased that comes to town a week later. Now we can do a LIVE broadcast from anywhere in our viewing area, which was previously blocked for microwave by the mountains. We do many live, multi-camera events around the United States including unmatched coverage of the Alabama-Auburn Football Games. We even covered live, a Huntsville Hockey team's championship game in Minnesota several years. Gus Hergert wanted to use it for sports events every chance he got, but of course, hourly cost and driving miles was a significant factor.

August 19, 1995 - The 14th annual "31 3-in-1 Triathlon" takes place at Old airport and Natatorium. We had been at first conducting, then sponsoring the event since 1981 when our staff put on the very first event.

August 25, 1995 - SHOALS STAR RADAR comes to the valley. This is a high-powered Enterprise Electronics Corporation (EEC) weather radar mounted in a Winnebago Motor Home. It is taken to the Shoals and put in service as an early warning radar for areas west to the Mississippi River. It has the Baron processor on it, which can see 16 levels of rain.

November 11, 1995 - Channel 31 announces the opening of THREE new area bureaus. They are located in Guntersville (Marshall County), Scottsboro (Jackson/DeKalb counties) and Fayetteville, TN (for Tennessee counties). This brings to FIVE, the area bureaus covering the whole 31 viewing area with regional/area bureaus for a local focus on news, un-matched by any other TV station.

September 7, 1995 - We also can get to News Events faster with AIRSTAR, a twin-turbine prop-jet MU-2 aircraft. It was a leased aircraft with a partnership and M.D. Smith, IV, being an instrument rated, multi-engine aircraft pilot, was also able to fly co-pilot on this fine, fast aircraft.

From my perspective, the nineties was the decade of the best years in the TV industry, even though the "Big Rollup" of TV stations by a few mega-corporations had begun. All the competing TV stations were owned by large corporate chains, and we were one of the very few "family owned" TV stations still around in a major market and one that was consistently beating all the competition for the ultra-valuable TV News audience. With number one ratings there, you tended to dominate other time periods as well, even the network times, no matter who was dominant across the U.S. It was because the top TV News operation generate respect and with much of the audience, if that station is the best at news, then their syndicated programs and their network must be the best also.

The networks knew the same thing, and our "hourly network rate" reflected the network's value of us in that position. And of course, local commercials cost the advertiser more because we were delivering more "eyes" to their commercials. Sometimes we even had a "premium" rate, meaning we could charge more per thousand people we had watching us just because we were number one, our availabilities of commercial time was at a premium and thus it simply cost more. That is the BIG reason all the stations, at least in our market, competed so hard for the news audience, and that was also to the benefit of the Huntsville and North Alabama viewers as well. When stations compete, everyone gets better and everyone benefits, even if some profits have to be used in that effort. Near the end of the decade, all the profits were required, and that is just ONE reason I started looking around to sell in 1999. But more on that later.

We did our very best to get and hold on to the best people we could find. Not only the on-air anchors, or even the on-air reporters, but every single position in the news department was of critical importance to us. The writers, the assignment editors, the photographers, the supervisors and certainly the News Director.

You see, when you have a top anchorman or anchorwoman, words gets around in other larger markets that pay more, and they can often "sway" a person to come to work for them with a larger pond to swim in besides a hefty increase in salary. Therefore, management from the top down did all we could to hang on to good people including some strong contracts and penalties for leaving early.

That was MY job. The buck sure stops at the top and my desk. I spent almost as much of my time learning how to be a better manager which included taking the Dale Carnegie Management Seminar back in the early seventies, and then went on to become instructor qualified to teach the course, and proceeded for the next nineteen years to conduct the Dale Carnegie Management course for the Birmingham and North Alabama franchise holder. I conducted the majority of evening courses, which was one weeknight for seven weeks, in mainly Birmingham and Huntsville. I also conducted seminars in Decatur, Athens and even down in Centre, Alabama. Every time I conducted a seminar, I'd work harder with my seven department heads to become better middle managers, learning better how to "build their people and make them successful," because the only way you as a manager can be successful, is for your people to be successful. That means building morale, team spirit and certainly motivation on the part of each team member. Not unlike a top football team. Talent is necessary, but great leadership is also required to be the best.

When a key player left, such as a News Director or a key Prime Time Anchor, it was a tough break, but we had to find a great potential replacement, and then start work on building him or her and make them successful. That applied to every department such as sales, engineering, production, commercials, art and let's not forget traffic and book-keeping. Every single job and the team spirit of the whole is important.

With a family-owned station and me at the top, I was able to work hard to do just that. My office door was always open. I always wanted to hear when an employee had a problem that they wanted to bring to me instead of or in addition to their department head boss. I never undercut the authority of my department heads, and sometimes against my better judgment let them keep an employee, and I felt they should be fired, but more importantly, I let them fire and very good on-air person because of the "trouble" they made behind the scenes. Usually, I didn't feel the proper time had been taken, meetings with quarreling employees were exhausted. There are many ways to solve problems between workers.

Just as there are many ways to solve problems between department employees. For example, the news cameramen might be loudly complaining to the engineering department about a repeated failure of a portable video tape recorder in the field, and the engineer is barking back that "if he were not so hard to that piece of gear, it wouldn't fail again in the field." That's the kind of things TV department managers face every day and keeping harmony between arguing employees and a key part of manager's jobs. There were occasions when I had to sit quietly in the conference room with the two department heads to try to work things out, and on occasions, we would also have the quarreling employees in there with us, seeking a better mutual understanding of each other's motivations, short tempers, and the fact that everyone wanted to do the best they could do towards a winning result for their department, the station and themselves. Sometimes it's only a matter of how it's *splained* to everyone.

You might wonder what 175 employees would do at our peak of employment. We were trying new ways to make money, new fields to branch out and you don't know until you try. For example, after a brainstorming session, we came up with the idea of having a "wedding and party video production" branch. We already had several good "commercial" photographers that are different from news photographers. Those people primarily go with a production man and perhaps the salesman, to the field to make video commercials for sponsors. They were on film in the early days, but by now, all color video cameras, some even with a small built-in video recorder that worked very well. We decided it would take three more to do this on a nearly full-time basis and we launched that branch. We advertised it on the air, and we did get a few jobs, and we did a wedding package unmatched, but it was not cheap. Most people preferred a friend or relative with a home movie camera and VHS tape deck to record the wedding, usually without any editing, titles or the production values we included. That department lost a lot of money, so after a year, we shut it down. We had to let those workers go, and I always hated to let good people go. It was not long after this year that we had to have a more general layoff, trimming 25 employees from the staff, but even so, we still had the largest staff of any TV station in Huntsville. They had been at gradual layoffs long before we started. It was a sign of times to come.

I had always believed that you have to try some new and different things if you want to stay ahead. That does not mean everything you try will be golden. You make mistakes, and you both learn and recover from them. If you get to be Number One in the ratings and try to rest on your laurels, you start slipping immediately. You are never able to remain "status quo."

It's hard to write about how to keep employees highly motivated and producing top results in such a variety of types of jobs as you find in a typical TV station, without discussing managing people.

I have touched on this topic already, having learned that lesson in the early seventies, after a stint of being a "slave-driver" of a boss at my father's direction in the sixties. I "felt" this was wrong, but I was young, and he not only owned the station, but had been bossing people most of his life, and had a good track record of doing it his way. It just didn't work for me.

I found out a "boss" is a person who gets behind people and cracks the whip like you might with a team of horses. People may do what you require and demand with threat of punitive measures for not doing what you said, but it's a short-lived success. Turnover and poor morale is the immediate indicators of such a management style. The one that works is the "leader" who gets out front and says, "Come on, follow me, I will lead you to success and give you credit when we all succeed."

I had posted in my office and asked my managers to have in their office a sign that says, "Good Management is the ability to build people and make THEM successful." Also, every good boss knows to "Praise in Public, and Criticize in Private." You should also try to make your praise public and often, but make sure it is genuine with specifics about what the praise is about, not just "Good job," and walking away. At our monthly full station employee meetings, we had several types of recognition awards passed out, and the one I liked the best and gave out myself was the **"Atta-Boy"** award. I took nominations from each department head, and I selected the winner. I still love that term to this day.

Programming a TV station in the Nineties was different in some ways and similar in others to previous years. For the syndicated, off-network shows such as "Bewitched" that we ran successfully for a number of years at the 6:30 p.m. time slot, to the hit show "Happy Days" which also ran successfully in that time period, we knew we had to pay top dollar to get it, AND be the first to get to the syndicator and dicker for the best deal without losing it to the competition. On a few rare occasions, it got into the bidding war.

That was one reason, I got to be VERY good friends with the traveling syndicated sales people that would come to the market a couple of times a year. I did not treat them like a typical salesman selling light bulbs, because I knew the relationship formed and bonds would get me a call first when a hot property off network from ten years of successful runs would be available.

Network times program themselves. It's anything the network chooses to run. You have the option of NOT running a certain show. For example, we chose not to run the first year of "NYPD Blue" because of the nudity and profanity it contained. That was a "first" for broadcast television, even though HBO and Showtime had been airing R-rated movies for a long time on cable. We didn't get paid either, had complaints from both sides from those who wanted to see the show and hated any form of censorship, to the other side when we decided to pick it up in year two since it had been running on the Independent TV station in the market. We had not stopped it from airing, just not on our TV station. Overall, it was a good move on my part since the proponents of the show could still see it, and the most conservative segment of our audience loved us for NOT airing such a lewd show. I personally liked it fine, and it was a well-acted, well-scripted series. My reason was for concern over what our audience wanted, and the feeling seemed to be split, so I erred on the caution side that time. I didn't always choose that path.

News Programming is a quite different segment of programming, it's the most vital, most important and makes the most money for a TV station.

You have every single commercial break as yours to sell compared to only station breaks of sixty seconds between network shows, and syndicated, you traditionally had three different two-minute breaks in the show and a seventy-second break at the end, for another minute commercial and a ten-second station ID.

In the sixties, all commercials were pretty much 60-seconds long, but by the nineties, almost all were only 30-seconds long and some only 15-seconds. Breaks were often expanded to two-and-a-half minutes long to make more money, while the audience suffered through at least five 30-second commercials and sometimes four 30s and two 15s for a total of six commercials in a single break. But money was getting tighter in the sixties, network revenue was falling off, and would eventually cease completely, and finally, networks began to charge for their programming like syndicated programs always did, but we didn't have all the breaks inside the network shows to sell.

So, NEWS was the savior. It could make or break your station's revenue. If you were number one, at a high cost per commercial, and ALL the spots in a newscast were yours to sell, it's the largest money-maker you have in any single half-hour show, and the morning news had by now expanded to start at 4:30 a.m. and run until the network morning show at 7:00 a.m.

Early morning didn't bring in as much per commercial in dollars, but there were a LOT of spots to sell, compared to just the 30-minutes at either 5:00 or 6:00 p.m.. It was the "Cash Cow" as we used to call it. It's probably still referred to that way.

Frank N. Magid Company:

We used the News Consulting firm of Frank N. Magid that we contracted in 1974. We had them under exclusive contract for our station in our market continuously until I sold the station. That was one of the wisest things I ever did. Both I and all the different news directors we had over the years, were guided by the market research they did, and their expertise of sharing top techniques of markets much large than us and what they did, Magid showed us what special news segments and features they had come up with, and we adapted our own "franchises" like "The Country Rover" featuring Jamie Cooper. That was a news director and Magid's combined idea that was a major hit for many years.

I had just begun to fully understand the "News Audience" under McHugh-Hoffman, but the Magid years were an education in themselves. Magid was primarily a research company and did research for some of not only the largest TV stations I the country as well as ONE of the major networks, whichever one had them under exclusive contract at the time. They made ABC's GMA, Good Morning America the #1 morning TV network show under David Hartman and others. And when they felt some years later, they didn't need Magid. Frank's company went to work for NBC and the "Today" show. Subsequently, it became the number one morning show.

Once you had the research to prove, what they could view was happening on all three stations and with 90% accuracy guess anyway, they would provide the Consultant services on a year or multi-year contract basis. Through the research and implementation told us what and who viewers liked and didn't like. Once you made some major and minor changes in both talent, leadership and the on-air product, you had to make the audience aware of it. They advised you to avoid **"an-ti-ci-point-ment"** with viewers. Don't "blow your own horn" about a coming new anchor person who will be the greatest anchor the market has ever seen and the public has great "anticipation" to the extent no one could match up to the hype, and then have the market "disappointed" because their expectations were set so high. That one coined word by Magid, I used a great number of times over the years because it said so much in just that one word.

I learned directly from Frank Magid himself, at lunch in Cedar Rapids, once when we leaders and news director flew there, the story of "The New Car." As he started, I wondered, "what in the world does a new car have to do with a favorite news station and promotion?"

Frank Magid explained. "When a man is in the market for a new car, because his current one is old and shabby, he starts looking around at ads for new cars. He watches TV commercials and visits new car dealers. He reads 'Consumer Reports' magazine. Then, perhaps after a few test drives in his narrowed down choices, he finally picks one, pays for it and starts driving his new car. His choice above all others and he is very happy with his choice.

"But then a funny thing happens. Not only does he stop shopping for any kind of car, he stops paying any attention to the TV ads in front of his eyes. He may think about something else, change channels, but surely it will be ignored by him. It will be that way for a long time. That is because he has MADE his choice and he is happy with it. He does not even want to know about any other features of any other car that might take the shine and glory off of HIS choice. He does not want to be unhappy or dis-satisfied in any way. If he paid attention to the ad and saw some new feature, he might start to become that way.

"Now, here is the parallel. The same thing happens when a person chooses one of the TV stations to become their favorite. They tell others that it is THE BEST. He/she knows why he likes everything about their news casts. He likes the stories the pick, the reporters they use, the anchor talent including sports and weather and for a period of time, will not even watch any of the competition. He/she is loyal, and he/she is not going to change for anything. But just like the car that gets older with time, it gets a few dents and scratches on it. Like his number one TV station, is coasting on their top ratings. The TV station has some turn over in talent which is inevitable to happen, the station quits trying so hard to beat the other stations to news and weather events and cover it markedly better than anyone else. So that station is becoming just like that aging car (if they don't do something almost continuously to "spruce" things up), showing its age. And if all else stays the same, eventually that man will start paying attention to the new car ads again. Just like the TV viewer will start paying attention to the 'promotions' and ads from the competing TV stations again. And you see, the cycle about to complete itself again.

"That's why TV stations have to keep improving themselves and exceeding their own standards and goals. If they don't, they are slipping behind. That is also why it can take years to become #1 from a #3 starting place."

That is an example of the sorts of things I learned over many years. We continuously kept the Frank Magid company under contract until I sold the station in 1999. Besides that, I attended every single meeting the visiting news consultants had with the news director and was part of every delivery of research study delivery and presentation of recommendations. Because if the News Director left, and that happened quite a few times, he was not the only one with that knowledge he was taking with him. I would work with and tutor the new News Director, sometimes send him to Cedar Rapids to train under Magid and learn why their advice was to be trusted and adhered to so strongly.

So that is just a brushing over the high spots of programming and news in particular. That's what it had become in 1995. It was "one hell of a horse race," as we used to say. And when stations compete as hard as they can for news ratings, all the stations get better and the viewing audience gets the benefit to choose from among the best. That was not the case in some much bigger markets like Birmingham, until much more recently. We were in locked battle primarily with Channel 19 since 1971 when we got our first news consultants of the McHugh-Hoffman firm and doubled our ratings in one year. We were still behind 19 in 1972, but the gap was closing.

In 1970 Channel 19 had Hans Sitarz as the lone primary news anchor and H.D. Bagley as a dry and homey Meteorologist, but he knew his stuff. And they were whipping our butts with just our nameless two news men and two photographers for the entire news department, and Johnny Evans doing the weather and a part-time guy doing sports.

We were the first station in 1972 to have Co-Anchors, a man and a woman, and with all the other changes we made over the coming next few years, took the Number One title away from Channel 19 in 1974, and really shook up their boat. It was not long at all that before Hans was gone and a pair of Co-Anchors replaced him on Channel 19. The rest of the story and the continuing news battles is history. In the early 90s, Channel 48 joined the race seriously and by the mid to later 90s, it was a three-way race to either get or stay on top.

1999 and Selling the Station

By 1998, the station was losing money as viewers were declining for all TV stations, and Channel 48 had made the news ratings war a three-way race. We had already found it necessary to reduce the staff from our all-time high of 175 to about 135 people. It was very tough to require my seven department heads to reduce their proportional amount, save for the sales department. It had to be done, or our losses would be overwhelming. That got us back on the right side of the bottom line. But a large expenditure in the range of three million dollars was looming to convert the station from the traditional analog to full digital in the next few years. And like most things, that estimated price for conversion was likely too low. Up until then, I had never even considered selling the TV station. My mother, having died in 1997 but being pretty much out of the picture, I had become the CEO and I just could not imagine how we could continue to serve the viewers of the North Alabama and Tennessee Valley the way we always had, with the best people, and some of the highest paid staff in the entire market.

About then, a broker named Richard Foreman happened to contact me about selling the TV station. His timing was perfect, and I told him, being a family business, we didn't have a huge profit line, as we usually plowed most of it back into the business, but I would sell for 3.5 times the gross revenue if he could find a buyer at that computed price. In one or two months, he found one that was a recently formed conglomerate buying up several middle size market TV stations. I'll spare you the thorny details of the "contract" and "due diligence" which was gruesome for both sides to find agreement, but we did. The final contracts, signed in Atlanta in a huge Law Firm's office, was FIVE inches thick. There were five sets, and I had to sign in multiple places on each set. The deal was done, announced first to staff, then to the public and the wait of about four months as the FCC has to review and approve any broadcast property sale.

In the meantime, I decided to give every single member of the staff a parting bonus from Smith Broadcasting, taking one million dollars of the sale, and dividing it up among the staff. Our formula was based on each employee's yearly income AND the number of years spent working at the station. Folks like Cactus who had been with me since

1963 and was in a top manager's pay scale did quite well. Recently employed hourly workers, not so much. We were a family, and to me, it was the only thing I could imagine doing, so I did. The day new ownership took over, the bonus checks were handed out along with a termination notice to every single employee.

Then the new owners offered employment contracts to the bulk of everyone on the staff, most at the same pay grade. A few not offered new contracts were pretty much told that in advance, and some had already found new jobs. All of my managers were replaced with the new company's team. We all expected that. And then it was over. I had already moved to an office downtown and supervised our remaining business, HiWAAY Internet Services, until it was sold in 2011. I thought it was wise to keep a budding technology company, but it proved disastrous to our corporation and me in the coming years, and we lost about all of the profit the TV sale had generated, after taxes.

That's another story for another time.

TV Post 1999, After WAAY-31-TV Sold

I have come to know the new General Managers that have been through the doors of Channel 31 in the nineteen years it's been since we sold the station. With declining advertising revenues and a debt load of the sales price of the station, it's been "more than hard" on the new owners, and the local GMs (General Managers) sometimes expected to deliver the impossible. You must keep revenues up, keep expenses down or even cut expenses, including salaries, increase viewing share in the market, and end with a profit large enough to pay off the debt load. Finally, have a bit of profit left over. I would hate to be assigned that task.

One previous employee from '74 to '77, Hank Price, has remained in the business and as this section is written in late 2018, he is the GM of Channel 13 in Birmingham, purchased by Hearst Broadcasting about three years ago.

(BY ALL MEANS, READ HANK'S STORIES IN ANOTHER SECTION OF THIS BOOK TO GET A FEEL FOR HIM AND SOME THINGS HE WROTE TO ME IN YEARS PAST HIS TENURE AT 31.)

I have kept up with Hank over the years. He has always been a good friend, and we share much in common. Just to kick it off, here's his work history in a nutshell.

WJTV Jackson, MS 1969-1973

WAAY 1974-1977

WPEC West Palm 1978

WPTV West Palm 1979

Frank N Magid 1980-1983

WJLA Washington 1983

WUSA Washington 1984-1987

WFMY Greensboro NC 1988-1991 (GM)

KARE Minneapolis 1991-1996 (GM)

WBBM Chicago 1996-2000 (GM)

WXII Winston-Salem 2001-2014 (GM)

WVTM Birmingham 2015-2018 (GM)

Notice he was at our station for three years, and because of his talent and ambition, he went to a better job at more money, and I wished him well. Notice also; he worked directly for Frank Magid Company ('80-'83) for three years as a News Consultant. He learned, like I did, those folks at Magid are the best we have ever encountered at research, then advising a station how to get to Number One and even harder, how to stay there.

His talent, drive, creativity and learning ability eminently qualified him to be a successful GM for whomever he might go to work. The last four stations (maybe five) were owned by the Hearst Corporation and he was brought in to "turn the station around" from bottom place to Number One or doggone close. He has done that. He has written several books on that very topic, aimed at a very limited audience of other TV General Managers and Owners.

Hank has also, by way of some phone calls we've had over the years and several in-person visits after he moved to Birmingham, told me of the drastic shape of TV in 2018 with the "Internet Age and Digital." It has changed the "game" like nothing I could have imagined in 1999. At that time, the ABC network, like other networks, was reducing the rate they paid stations for carrying their programs and all the commercials in them that they were delivering to viewers in the local markets. Now, it has been a reversal, and TV station pay the Networks great sums of money every month, much of which comes from "Re-Trans Fees." That is what the cable companies pay local TV stations to "re-transmit" their local news and network TV shows on their cable systems. The networks get a large share of that pie.

You have not only those kinds of financial changes, but young people don't watch a thirty-minute newscast anymore, if they ever did. They want "news on demand" and usually in no more than 5-minute-chunks. They rely on their phones and tablets for that. So, TV stations have to compete for those eyes, and the station's web site must have all kinds of news archived, searchable and easily found, for those who only want to hear one story. You don't want them going to your competition, so you have to supply it before someone else does.

At the same time, you still have a regular bunch of news shows, more than ever in terms of hours, beginning as early as 4:00 in the afternoons up until 6:30 (for Central Time Zones) and also starting very early morning until 7:00 a.m. when network programming starts. All of the reporters "package reports" and now generally shot on digital cameras, edited digitally, and besides air use, may go straight to the web site, mobile TV platform, and a host of other ways it may be used.

It astounds me where the media is today. Many people even rely primarily on platforms like Facebook for the bulk of their news, and much of it may be fake, which makes matters even worse. No wonder the country is so polarized these days.

Well, I take my hat off to the GMs around the country, like Hank Price, who are still persevering and making their stations number one in the market and increasing revenues, which is "one hell of a balancing act." Where is the TV business going in the future? Only the most creative and innovative stations will continue to survive. Others will go into bankruptcy and perhaps even close their doors, like so many AM Radio stations around the country, have already done.

If you don't stay with or ahead of the leaders, just wait to see what happens. A parallel example I think, is Amazon, making it so easy and effortless to shop at home, and usually free shipping. Walmart and Target stores are hot on the heels of building their on-line shopping and mailing departments and currently, it seems doing a good job of it. Sears, on the other hand, the "Catalog Sales Leader" for over One Hundred Years, is headed out of business.

People and businesses resist change, but leaders know that it's not only inevitable, but it's also demanded. A little mail order music company of a few years ago became Amazon. Changes are happening at "light speed." To make matter worse, the RATE of change is also increasing, exponentially at that.

What a ride it was for me from 1963 to 1999 in television. But it was nothing compared to what GMs the past nineteen years have experienced. Furthermore, the future rides are truly unimaginable.

The End # #

About the Author

M.D. (Marcellus Davidson Smith, IV) was born in December 1940 in Birmingham a year before Pearl Harbor and remembers some of the "blackouts" of WWII at nearly age five, while listening to radio.

He listened to Radio Mystery Dramas with his father as a kid. Radio and later Television has always been a part of his life, like breathing. From the Friday Night Fights on radio sponsored by Gillette Blue Blades, to all the mystery programs like "The Shadow" and "Inner Sanctum."

Hearing WBRC-FM shortly after it went on the air in the late forties, near Tampa, Florida one evening on vacation on a new Zenith AM-FM receiver was a thrill to find out it would carry that far.

M.D. who intended to be a doctor like his father (with an "M.D." at either end of his name) always said he'd like to have been, changed from Pre-Med to major in Broadcasting in 1960. He married Judy Chandler in 1961.

In 1962, he got his Amateur Radio License, WA4DXP, and has been on the air with "ham" radio off and on ever since. He is currently active, favoring digital modes for very long-distance work.

After graduating in 1963 from the University of Alabama with a two-month-old son, he went into the family broadcasting AM Radio business, joined shortly by the first TV station where he continued to work for the next 36 years until he sold the TV station in 1999. M.D. III, as his father was called, and M.D. IV, learned flying in 1965 and took many trips over the next 20 years, most in several models of a Piper Twin Aztec.

He has been writing short non-fiction stories for thirty years, with over fifty of them published in several different publications.

Made in United States
North Haven, CT
20 October 2026

20740141R00186